C000065248

A MANUAL

OF THE

SEA-ANEMONES

COMMONLY FOUND ON THE ENGLISH COAST.

BY

THE REV. GEORGE TUGWELL,

ORIEL COLLEGE, OXFORD.

LONDON:

JOHN VAN VOORST, PATERNOSTER ROW.

M.DCCC.LVI.

LONDON:

E. NEWMAN, PRINTER, 9, DEVONSHIRE STREET, BISHOPSGATE STREET.

This scarce antiquarian book is included in our special *Legacy Reprint Series.* In the interest of creating a more extensive selection of rare historical book reprints, we have chosen to reproduce this title even though it may possibly have occasional imperfections such as missing and blurred pages, missing text, poor pictures, markings, dark backgrounds and other reproduction issues beyond our control. Because this work is culturally important, we have made it available as a part of our commitment to protecting, preserving and promoting the world's literature. Thank you for your understanding.

TO

WILLIAM BRODRICK, ESQ.,

ONE OF THE AUTHORS OF 'FALCONRY IN THE BRITISH ISLES,'

FROM WHOSE PENCIL

THE ILLUSTRATIONS OF THIS VOLUME ORIGINATED,

THE ENSUING PAGES ARE INSCRIBED,

AS A SLIGHT TOKEN OF ESTEEM AND GRATITUDE,

BY HIS FRIEND,

THE AUTHOR.

"Are not the mountains, waves, and skies, a part
Of me and of my soul, as I of them?
Is not the love of these deep in my heart
With a pure passion?"

CHILDE HAROLD.

"Mighty Earth,
From sea and mountain, city and wilderness,
In vesper low, or joyous orison,
Lifts still its solemn voice."

ALASTOR.

"Benedicite ... omnia quæ moventur in aquis
DOMINO; laudate et superexultate Eum in sæculo."

CONTENTS.

LIST OF PLATES.

INTRODUCTION.

Let me premise, by way of introduction, that the ensuing pages lay no claim to the character of a scientific treatise on the subject of Marine Zoology, nor are they a monograph on the genus indicated by the title-page. They are not, therefore, addressed to professed naturalists, but to that section of amateur ramblers about our English coasts who take a pleasure in noticing every form of beauty which they may encounter in their wanderings. And I am desirous of saying a few plain words about sea-anemones to such an audience for the following reasons:—

In the first place, the sea-anemones are objects of exceeding beauty, both of form and colour; they have always been compared (and do not suffer by the comparison) with the most gorgeous and delicate flowers of our woods and gardens—with the frail wind-crumpled anemonĕ—the stately dahlia—and the many-tinted chrysanthemum. Poets have sung their praises, and strong-minded naturalists in speaking of them have for the nonce risen above the shackled routine of their ordinary scientific descriptions.

Next, they may be obtained by a very small amount of labour and research: there are few rock-tracks between the tide-levels which will not afford a harvest to an explorer of ordinary observation.

Further, they may be kept in the drawing-room or the study without trouble or annoyance, and may be carried away to our country-houses as a permanent memorial of our visit to the sea-side.

So, too, they form an excellent means of introduction to the study of Natural History. Few people have either the time or the inclination to enter upon the vast field of the animal and vegetable kingdoms, with the intention of mastering either subject as a whole. Few people, for instance, could be induced to attack a book like Carpenter's 'Principles of Physiology,' in which the general laws of the subject are scientifically detailed. But very many persuade themselves to invest a little spare money in a Handbook of Ferns, or a popular treatise on Butterflies, and after a time discover to their delight that a small expenditure of time and observation has enabled them to become familiar with one branch of the great subject. Thus they are led on insensibly to become Natural-History students, and find at length that they have made no small progress in the journey which they feared to commence.

It is impossible to identify the commonest of sea-anemones without gaining by the way a portion of scientific knowledge; we become, by the very process, scientific observers; we get a little information about the construction, or the anatomy, of the animal in question, and therefore of all animals; we get into a habit of discrimination, of noticing what sort of formation and of habit is peculiar to a sea-anemone, and what those characters are which it shares in common with certain other animals: we therefore analyze, generalize and classify—very hard words indeed, but the signs of very simple thoughts.

I need not add that such processes strengthen the memory, improve the reasoning powers, and raise the mind to an habitual love of the great Maker of these varied and wonderful creations.

From these motives, and they seem to me sufficiently good and reasonable, I have selected the class of animals generally known as sea-anemones, and intend to describe as simply and clearly as possible what they are like, and where they may be found—to tell my audience (presupposing, as an author must, that there be an audience) a few facts about their structure, their habits and their classification—and to add a few hints on the best method both of discovering them and of keeping them, though in captivity, yet in a state of natural health and beauty. I do not mean to speak of every individual of the genus, but chiefly of those which are more generally found on our shores, and I have laid the scene of my book on the coasts of North Devon, because it was there that I first learnt to delight in this fascinating pursuit.

In conclusion, I must repeat that this Manual is intended only for the unlearned; that its language, and possibly its classification, will not be scientifically irreproachable; and that its highest object will be attained if it induce a single mind to take an interest and delight in the pursuits of Natural History, or even in the study of so simple an object as a

SEA-ANEMONE.

Ilfracombe,
February, 1856.

A Manual of Sea-Anemones.

CHAPTER I.

WHAT IS A SEA-ANEMONE?

EVIDENTLY this is the first point to be settled; but, though the question is a simple one, I am afraid it will take more than a few words to frame an intelligent and intelligible answer.

Suppose we take a morning stroll on Capstone, and ask the first half-dozen of our friends whom we chance to meet, "What *is* a sea-anemone?"

"A sea-anemone? Oh!—um—well—I'm sure I don't know!" That is a candid answer at all events.

"A sea-anemone is a gelatinous animal, which is found at low-tide adhering firmly by its base to the rocks." True, perhaps, so far as it goes, but not satisfactorily distinctive; we are rather afraid that our friend is one of the scientific humbugs—animals which are also found in great numbers about the rocks at low-water.

"Oh! *I* can tell you—it's the prettiest thing I ever saw in all my life—it opens and shuts just like a flower—and it's all over the most be—a—utiful colours you can possibly imagine." We are of course much indebted to our young-lady friend for her obliging remarks, but cannot help wishing that her enthusiasm had induced her to institute more accurate observations upon the animal in question.

Our remaining friends all appear either to have "read" or to have "heard" about sea-anemones, but do not seem to have benefited by either process. One person alludes to them as "interesting proofs of the natural adaptations of means to an end," but does not further explain himself. Another, having puzzled his brains with reading a popular account of the distinctive characters of the vegetable and animal kingdoms, has a vague idea that they are connected in some mysterious way with sea-weeds; and the last, being of a practical turn of mind, enquires how you can ask him such a foolish question—of course he knows what a sea-anemone is—it's a fish (which, by the way, it *is'nt*)—but, for his part, he can't see the use of your dabbling and splashing in muddy puddles, cutting holes in your shoes, and getting your feet wet, when it's so much more easy and agreeable to walk up and down the promenade, and see one's friends, and hear the band play. As there is plainly no chance of our agreeing with this gentleman, we wish him good morning, and stroll homewards, feeling that the collective knowledge we have obtained from our half-dozen acquaintances is by no means an equivalent to the trouble we have incurred in obtaining it.

Now, perhaps, as the reader has expended a portion of his patrimony in the purchase of this volume, he will begin to grumble when he gets so far as page the seventh, and finds that the writer has hitherto ingeniously contrived to shirk the point which he promised at starting to elucidate. So let me open a few books of reference, and turn up for my ill-used friend the best description that I am able to discover.

"A sea-anemone," I find it said, "is a radiate animal—an actiniform polyp. Body single, fleshy, conoid, fixed by its base. Locomotive. Mouth in centre of upper disc, surrounded by one or more series of conical, tubular, retractile tentacula."

There—what do you think of that? Supposing you had never seen a sea-anemone, do you believe you could go down instantly to the rocks, and bring back a specimen or two without any difficulty?

Well, as I am sure I could not in such a case, I shall proceed to say a few words about animals, and explain (in English) why an anemone is called a radiate, a polyp, and an actinoid—and then I think it will be extremely evident *what he is*.

Now the reader may think that this is a very long process, but it is positively a necessary one; and I trust if he is getting weary of long words he will skip the rest of this chapter for the present, and return to it when he has finished chapter the second.

Let us "begin at the beginning," and as one of our friends has asserted that a sea-anemone is a sea-weed, *i. e.* a vegetable, we shall convict him, in the first place, of an unintentional error.

All objects which exist on the earth's surface, and whose existence our senses enable us to perceive and recognise, are divided into two great groups, which we call respectively *Organic and Inorganic bodies.*

Now, what we mean when we say that we divide any given number of objects into two or more groups, classes, &c., is simply this, that some particular character or resemblance is peculiar to all the objects under each division respectively, which is not shared by the objects included under the other division—as, *e.g.* if we divide all men into good and bad, we merely intend to say that certain moral qualities belong to all the individuals of the former class which are not to be found in any individuals existing in the latter — thus, then, the peculiar character of all inorganic bodies (viz. *minerals,* whose atoms have no *organization,* —that is, do not work together with the object of gathering nourishment from surrounding matter for the bodies of which they form a portion) — the distinctive character, I say, of such bodies is *rest;* they cannot change, except so far as other bodies may be joined to them by influences over which they themselves have no control. A stone continues a stone, a piece of lead must ever be a piece of lead, although it may increase — or *grow,* as Linnæus says—by the addition of other portions of that substance, but it, of itself, has nothing to do with such a proceeding. Now by this distinctive character we recognise

The Mineral Kingdom.

But *organic bodies,* on the other hand, possess *life,* that is, motion dependant upon themselves; thus, trees and animals increase by the sap and the blood which are circulated through themselves by means of the organs which are given to them by their Maker when they are formed.

Hence we get our first great division of all natural objects into

ORGANIC and INORGANIC bodies.

But a sea-anemone (I must presuppose for a short time that the reader has some slight acquaintance with the object) evidently increases by means which he possesses in himself, and therefore is not a mineral or an inorganic body: we will then leave this division to take care of itself, and proceed to the consideration of *organic bodies*, under which head our new acquaintance is, we find, included.

These organic bodies are divided into two portions called animals and vegetables, but we have seen that to form a legitimate division we must get a definite idea of more characters which are peculiar to each class: what then distinguishes an animal from a vegetable? You will say that common sense shows you the difference between a cow and a cabbage—it is impossible to confound them. Very well, but what does common sense say about a sponge? I have a work on Physiology, published only four years ago, in which it is stated that our most distinguished naturalists are still divided on this point. Now one mark by which we recognise an object as an animal, distinguishing it thereby from a vegetable, is the fact of its having a distinct cavity set apart for its organs, as the skull for the brain, the stomach for the digestive organs, the chest for the lungs—whilst in plants the respiration, for instance, that is the reception and rejection of the air necessary to their growth, is carried on by means of infinitely extended surfaces, represented by their leaves: so, too, animals generally grow by adding new matter to themselves, which only keeps up their definite form, whilst trees and flowers perpetually put out new branches and roots without any very definite limit. Again, vegetables feed on inorganic substances, lime and so forth, and animals feed on the organic* substances, as sugar and starch, which the vegetables produce. Again, vegetables consume that part of the air which is called by chemists carbonic acid gas (and which is destructive of animal life), and give out oxygen, but animals consume oxygen and give out carbonic acid. Lastly, animals have the power of voluntary motion, and possess sensation, *i. e.* they feel by means of peculiar organs, which are generally recognisable, and which we call their nervous system, and plants are destitute of both. Of course many exceptions will occur to the intelligent reader, but what I have stated are the general laws; and we therefore feel justified in adding a second division to our catalogue, and asserting that all organic bodies (bodies possessing the capability of increasing by their own intrinsic powers) are divided into two classes called

The ANIMAL and VEGETABLE KINGDOMS.

Now, you will plainly see that a sea-anemone is included in the former of these divisions—and therefore that it is *not* a sea-weed. But if it is an animal, may it not be a fish, as one of our friends has been good enough to inform us? Certainly not, and for these reasons:—

All those natural objects which we cannot include under the head of minerals and vegetables, and are consequently *animals*, are divided by naturalists

* "Organic," as having been produced by vital organs.

into four great classes, about which I shall say a very few plain words.

The first class includes those animals termed *Vertebrate.*

A *vertebra* is the proper name for those bony rings which, in great numbers, compose the back-bone or spine of a man or a fish, and with which we are sufficiently familiar in the pictures of skeletons, or in the remains of a salmon or herring at our dinner-tables. *Vertebrates*, then, are animals which possess a back-bone, and this back-bone is the channel through which the nervous centre is continued from the brain into the trunk, and from which nerves run to the different members of the animal's frame. But an anemone has no back-bone—and a fish has, as well as a man, this peculiar structure: therefore, plainly, an anemone is not a fish, and a fish is more nearly related than an anemone to a man.

The second division includes animals called *Articulate:* this word means jointed, and is applied to creatures whose bodies, like the earth-worm, the centipede or insects, are formed of a series of segments, and whose skeleton or hard parts (when they have any) are external. Their brain is very small, and is placed on the gullet, and thence one or two cords pass along the stomach, and, uniting little knots of nervous matter one to each ring, are continued throughout the length of the creature. Thus, if we cut some worms into two or three pieces, each piece continues to exist, having in itself sufficient "nerve" to move and get its living. But the characters of this class do not agree with those of our sea-anemone.

Thirdly, we come to the *Molluscs*, animals which do not possess a back-bone, or a jointed external skeleton. Their bodies are soft [*molluscus*, "soft," Latin] throughout; they have no limbs; their skin generally covers their bodies in a loosely-folded form, like a "mantle," which name it bears in works on Natural History—sometimes it is covered with a deposit of lime, and then becomes a shell; their blood is white and cold; their nerves are placed in masses not symmetrically disposed throughout the body. Examples of this class are snails, slugs, Trochi (white and pink, spiral shells often found among the rocks), oysters, cuttle-fish, &c.

Lastly, we have the *Radiates*. A radiate is a creature whose organs are arranged in *rays* around a centre, in which is usually the mouth. His nerves, when visible, are similarly disposed; and the stomach is equally developed on all sides of this centre. Plainly, then, friend Anemone is of this kind; for if you look at him when he is in full flower, or even partially open, you will observe his swelling lips in the midst of all his tentacles or long arms, which are busily engaged in collecting his food; and you will see that no one side of him is different from another

—that he is "rayed" or circular in every respect. So he is a *radiate animal,* and therefore not a vegetable, and not a fish. Very well; that is a great point, which we have gained without much trouble, I think. Now, I shall ask you to have a little patience, and go a few steps further.

We have spoken of the *Animal Kingdom,* and have arrived at its fourth division, the sub-kingdom of the *Radiates.*

These radiates are divided into several classes, one of which is that of the

Polyps, or True Zoophytes.*

The meaning of zoophyte is "living plant" (*zoon,* "an animal," *phyton,* "plant"—Greek), and the animals included in this class are so called, because, in the first place, they were for a long time considered to be vegetables; and because, secondly, a vast number of individuals are found united, like flowers on a plant, by a common stem. If you go down to the beach, and pick up the first object which you suppose to be a very delicate sea-weed, you will probably see (with a magnifying glass) that it is an assemblage of horny cells, or hollowed vessels, on a stem of similar structure; and, if the animal be alive, each cell is

* *Polyp* means "many-footed" (*polus,* "many," *pous,* "foot"—Greek), and the present class of animals bears this name in consequence of a fancied resemblance of the tentacles which surround their mouths to the limbs of an Octopus (*e. g.* cuttle-fish), called Polypus by the ancients.—Cuvier, vol. iii., p. 7; vol. iv., p. 430.

tenanted by a little creature of most beautiful form and most active habits. All polyps are not thus clustered, but many are, and the distinctive characters of the class are the facts of their being *fixed*, either solitarily or in masses, by a stem, and their possessing arms (tentacles, *e. g.*), with which they seize their food.

Other classes of Radiates are—

2. *Acalephs*, or "stinging animals," as "sea blubbers" or jelly fish, which swim freely in the waters by the alternate contraction and expansion of their body.

3. *Echinoderms*, or "prickly-skinned" animals, as "star-fish," "sea-urchins" and the like, also moving freely in the waters: with these we are not concerned, for our sea-anemone is a "polyp" or "true zoophyte," being an animal which is generally *fixed* to the rock which he chooses as a dwelling-place.

I must inflict one more series of divisions upon the student, before we can get a proper answer to our question.

These "polyps" are of three kinds:—

1. *Hydraforms*. Each polyp is solitary or joined to a stem common to a great many individuals; and the stem or "polyp-house," as it is called, is horny, and hollowed like a reed. Example, Hydra, Sertularia, &c. [*hydraform* means "hydra-like."]

2. *Asteroids*. The polyps are united in families like those of the last division, but the "polyp-house"

is fleshy, and the polyps open out on its surface like a star with six or eight rays [*asteroid* means "star-shaped."]

3. *Helianthoids.* The polyps are *single,* and attached temporarily or permanently to their dwelling-places in the rocks. Some few being "free," and some—*e. g.* corals, surrounded with a crust of hard lime—their tentacles (or arms) open out in a circular shape, like the rays of the sun [*helianthoid* means "sunflower-shaped."]

Now a sea-anemone is a Helianthoid, or sunflower-shaped polyp, and all Helianthoids are very near relations of our anemone. Let us, then, pause here, and draw up a brief catalogue or table of the divisions about which we have been talking, and we ought to get a very clear notion into our heads of the form and character of our friend.

* *All Natural Objects are divided into*

* ORGANIZED	and	UNORGANIZED BODIES.
e.g. man, beasts, trees.		*e.g.* stones, lead, iron.

[Bodies possessing, or not possessing, the means of adding to their own growth by their own implanted powers, exercised by definite portions of their bodies called organs.]

---

* The following Table is based on general laws, to all of which there are exceptions. The principles of classification are now undergoing investigation, and the present system may not be regarded as perfect or permanent.

* *Organized Bodies are divided into*

† THE ANIMAL KINGDOM	THE VEGETABLE KINGDOM
Feed on organic substances.	Feed on inorganic substances.

† *The Animal Kingdom is divided into*

VERTEBRATES.	ARTICULATES.	MOLLUSCS.	‡ RADIATES.
Animals possessing a backbone, containing a nervous cord, enlarged at its extremity into a mass called a brain, in a receptacle or skull. *e. g.* Man, fish, birds, reptiles.	Animals usually with a jointed external skeleton; nerves in knots disposed symmetrically throughout the body. *e.g.* Worms, centipedes, insects, crabs, &c.	No limbs, soft body; sometimes covered with shell; nerves in masses disposed irregularly throughout the body. *e. g.* Snails, oysters.	Organs usually arranged in *rays* round a centre. *e. g.* Sea-anemones, star-fish, sea-blubber, hydraform polyps, &c.

‡ *Radiates are divided into several Classes, among which are*

§ POLYPS, OR TRUE ZOOPHYTES.	ACALEPHS.	ECHINODERMS.
Animals fixed on stems, or solitary, with non-ciliated tentacles grasping their food. *e. g.* Sea-anemones, corals, Hydra, Sertulariæ, &c.	"Stinging-animals," moving freely in the sea. *e. g.* Sea-blubber.	"Prickly-skinned" animals, usually moving freely. *e. g.* Star-fish, Sea-urchins, &c.

§ *Polyps, or True Zoophytes, are divided into*

HYDRAFORMS.	ASTEROIDS.	HELIANTHOIDS.
Polyps single and associated; stomach without a distinct wall; reproduction external; tentacles variable in number. *e. g.* Hydra, Sertularia, &c.	Polyps associated, supported in a fleshy mass, or polypidom, stomach with distinct walls; tentacles in definite number, 6—8; reproduction internal. *e. g.* Sea-fan, Dead-men's fingers, Red coral, &c.	Polyps single or connected only by a creeping stem, free or attached; soft or encrusted with lime; stomach free, number of tentacles indefinite; reproduction internal. *e. g.* Sea-anemones, lamelliform corals, as Madrepores, &c.

Now, if you will begin with the few words in the last column, and thence carry your eye back, by means of the directing marks, to the columns headed respectively "Polyps," "Radiates," "Animal Kingdom," "Organized Bodies," you will arrive at an answer to our question—

"WHAT IS A SEA-ANEMONE?"

and I think the answer cannot be shortened, without running a great risk of vagueness and inaccuracy.

The intelligent reader may say, This is all pretty clear; but, in the first place, how am I to know an anemone from a coral or any other sunflower-shaped animal? and next, why do you use such hard Latin and Greek words, and break your promise of explaining your meaning in English?

Two grave accusations—one of inaccuracy and one of pedantry. Allow me to apologise and defend myself. The fact is that a "popular" book on a scientific subject is always in a dilemma. If it uses "hard words" and goes into minute details, it frequently ceases to be popular. If it does not select this severe course of action, its vague descriptions do not enable the reader to name the animal he has found, and give him no insight into its structure and character. When people write about animals, plants and the like, they use "hard words" for the following reasons: *words* are only the signs of *things*, and it is not advisable to use the same word as the sign of *two* things; therefore, when we get hold of a new plant or animal we have to find a word which has not been previously used, and we "coin" or make a word; and since there are botanists and naturalists in other lands besides our own, we use a word which they can pronounce and understand, and thus the Latin and Greek languages become a means of communication in all ages and all countries. Again, we get more accuracy by using a foreign language; for instance, one kind of anemone is called *Actinia alba*, which means "the white" sea-anemone; but if we used the English form it is evident that the term "white" must never be applied to any but this *one* anemone: two other species are white, but we could only call *one* so, for fear of confusion. Therefore, the fact is that if we want to understand the subject of Natural History, even to a slight extent, we *must* drive a few "hard words" into our heads. It will be very good for the memory; and will present very little difficulty, even to those who are not learned in the Latin and Greek tongues, if they take two simple precautions:—First, remember that a long "hard word" *means* something; Secondly, get some one to translate it or tell you what it means: you will then connect an *idea* with it, and this will aid your memory, and sometimes furnish you with a useful *fact*. So you will learn *things* and not *mere words*. For instance, take the word "Radiate;" at first you think there is no reason why it should not be "Chroton-hoton-thologos," or any other piece of nonsense, but once understand that it means "rayed," and you will learn a fact about many animals of the class, and will recollect the word itself without difficulty.

Next, as to the distinction between a coral and an anemone. I paused at the end of the *Helianthoids*, or "sunflower-shaped" zoophytes, because, after so many long words, we seemed to require a rest, and this was the most natural halting-place.

Now we will proceed, and separate our friend from his numerous relations. Having settled that the animal in question is a *Helianthoid*, we ask two questions:—

Is he partly covered with a hard shell or coating of lime, firmly fixed to the rock?

If so, he is a coral; and we pass on.

But is his body fleshy and covered with a skin harder than the rest of him?

If so, then he is *not* a coral, and we ask a few more questions about him:—

1. Are these single polyps found to be united, not on a common stem, but by a thin band of flesh at their base, something like the runners of strawberry plants?

If so he is a *Zo-anthus* ("live flower"), for a plate of which I may refer you to Mr. Gosse's new work on Marine Zoology.

2. Are these polyps entirely single, not joined in any way?

If so—

Are his tentacles (long feelers or arms surrounding his mouth) in tufts, like patches of mignonette round a flower-bed, or in circles without any break in them?

If the former be the case, the specimen is a *Lucernaria*, and *lucerna* means "a lamp," and therefore the name is (as usual) descriptive of these tentacles hanging round the creature, like lights in a chandelier.

If his tentacles are in regular circles, he belongs to the "family" of the *Actinoids*, or "sun-beams"—a name which forcibly recals the animal when you took him out of the cool depths of his rock-pool, and every tentacle shot out its many-coloured lights among the pink corallines and the dark tresses of the sea-weeds.

You will have a few more questions to ask, and we will (to save time) set them down in a tabular form, heading them by the questions we have already asked, and they will be always at hand for future reference.

Helianthoids ("Sunflower-shaped Zoophytes").

1. Is your specimen covered with a hard shell of lime adhering to the rock?

Then it is a *coral.*

2. Is its body fleshy and its skin firm?

Then further

A. Are the individual polyps united at the base by a fleshy band?

Then it is a *Zoanthus.*

B. Are the individuals quite single?

Then further—

a. Are the tentacles in tufts, at distant intervals?

Then it is a *Lucernaria.*

b. Are the tentacles in uninterrupted circles?

Then it is a "sunbeam" or an *Actinoid.*

If an *Actinoid*—

1. Are the tentacles without a hole or outlet at their extremity?

Then he is a *Capnea* or a *Corynactis.*

2. Do the tentacles possess this outlet, and also does the animal draw them up within his outer skin when he is disturbed?

If he does, he is not an *Anthea*, about which I shall shortly speak, as he is found in great numbers amongst his relations, the sea-anemones.

If he does so—

A. Is his base narrow and unfixed, after the fashion of a beetroot just unearthed?

Then he is an *Iluanthos*, and lives in the mud (*ilys*, "mud," *anthos*, "flower"—Greek).

B. Is his base broad, and the animal immoveable?

Then he is an *Adamsia*, and lives round the mouth of shells.

C. Is his base broad, and does he possess the power of moving about by means of his base?

Then, he is an *Actinia* or

A Sea-Anemone.

And if you cannot discover him now, as well as many of his relations, I am afraid that I am unable to offer you any further assistance.

CHAPTER II.

WHERE IS A SEA-ANEMONE TO BE FOUND?

Let us go down to the rocks together. It is a glorious afternoon in the early summer time. A cool sea-wind is blowing from the westward; and the vertical sun-blaze is quenched from time to time by solitary masses of soft white cloud majestically rolling in from Lundy, or dimmed by those delicately-barred and fringed troops of cirri which are sailing in the upper current of air from the far-off line of the Welsh Mountains. Yesterday a heavy ground-sea was surging in from the Atlantic, but now a scarcely perceptible rise and fall of the waveless tide is swirling among the distant peaks of rock, and playing with the sea-weed tangles, as a strong man with the glistening tresses of the wife of his heart.

The tides are at their "spring," with a fall of two-and-thirty feet, and another hour will bring us to the flood—what more, then, can a naturalist desire?

Let us go. Suppose we leave the Promenade and the Tunnels to our friends—especially him of the "practical" mind—and climb yonder range of hills, where seven Torrs, like seven fair jewels in a king's crown, sun-emblazoned, beautiful, girdle this pleasant valley, and hush the din of the shore-breakers on stormy winter nights. Across the fern-hidden, wandering, many-voiced Wilder. Past the hazles and the hawthorns, and the meadow-grass, where the corn-crake shrills in the land, day and night, his dry and carking ditty. Under the furze copse, where the heavy-scented glories of its golden blossoms are gleaming, where the crisp purple heather and the climbing scarlet tangles of the dodder and the fresh green volutes of the young fern-leaves, yield a home and a happy "pleasaunce" to the insects, and the birds, and the countless, restless troops of the rabbits, who, among the well-known mazes, hold perpetual holiday. Another step, and we stand on the verge of a precipice, and look down upon the grey rocks, a hundred feet beneath us, and faintly hear the quiet breathings of the sunlit sea. We will follow this sheep-track, which winds round the edge of the cliff—a dangerous path enough on winter evenings when a heavy gale is blowing from the westward, and the long Atlantic rollers are breaking in foam-clouds on the shore. But there is no hazard on this quiet afternoon, so onwards, rapidly and fearlessly; and now we descend the triangular slope of Torr Point, carpeted with the slippery, shining grass of the sea-thrift, and fringed with the white campion blossoms and the salt foliage of the samphire. Here then let us pause; and whilst we rest on this lichen-covered rock, which forms the weather-beaten base of the Point, we will settle the plan of our campaign against the Anemones, and talk over the arms and ammunition which we have brought with us for our expedition.

We have seen that an anemone is a soft-bodied animal, sticking firmly by his base to the rocks, and when we add that he is usually covered by the sea, we shall perceive that we want, first, something to detach him from his lurking-place without injuring him, and, next, some means of carrying him home with as little shock to his constitution as possible. With regard to the first point, let me draw your attention to this leathern case (plate 11, fig. 5), which is capable of being slung round your shoulders by a strap. It is 10½ inches in length, 3 in width and 1¼ in depth, and contains a double-headed hammer, a long chisel, an oyster-knife, a putty-knife, with a round point, an old ivory paper-knife, and a small net, made by twisting a piece of brass wire into the shape of a circle with a tail to it, and fastening a bag of muslin round the edge of the ring. As to the hammer and chisel, these are indispensable; a great number of the anemones delight in rock-holes, and it is impossible to get them out without chiseling away

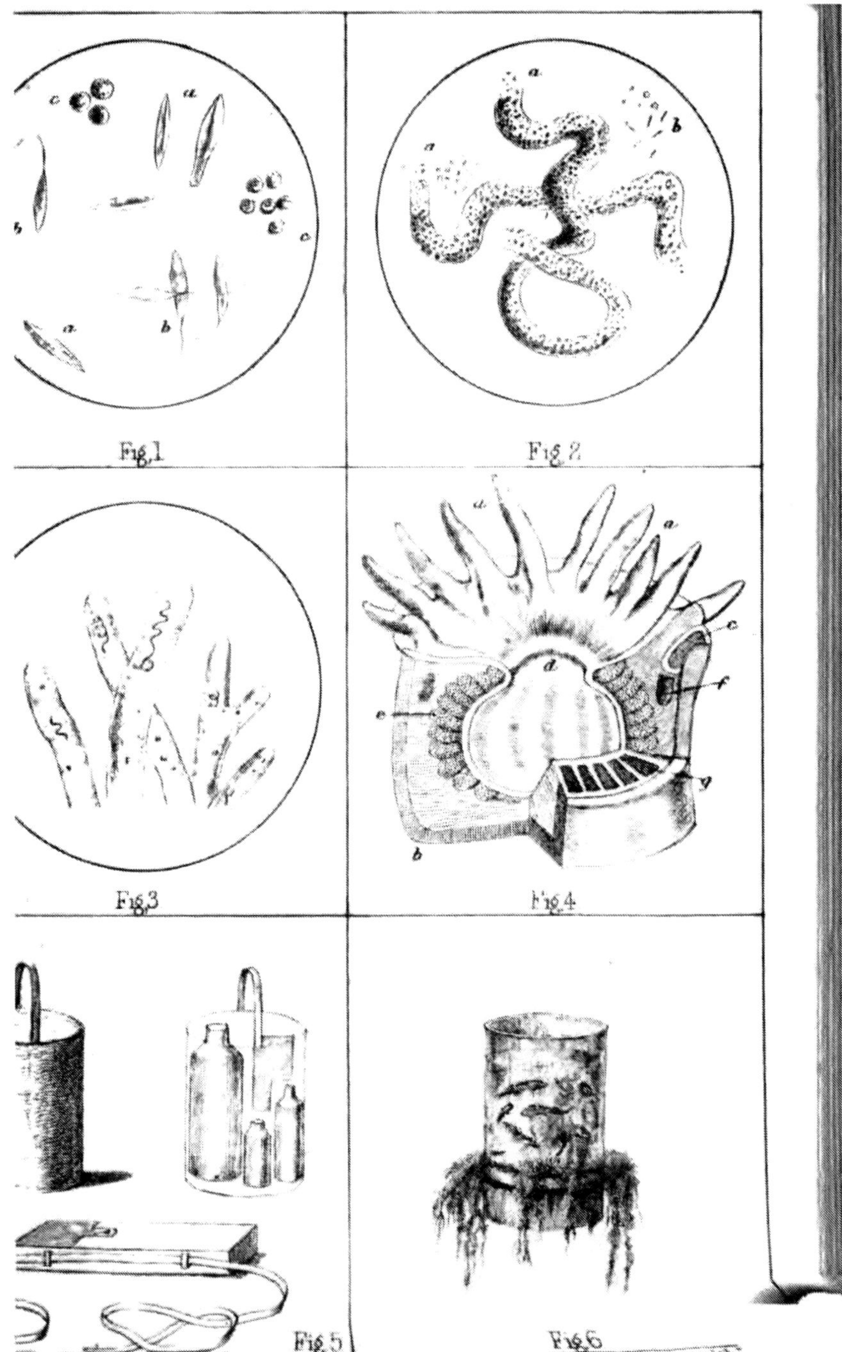
a
b
c
d
e
f
g
Fig.1
Fig.2
Fig.3
Fig.4
Fig.5
Fig.6

a portion of the rock to which they adhere: we shall see presently that "the daisy" will shrink to the size of a bean on being touched, and retire into so deep a cavity that no knife-persuasion can move him. The oyster and putty knives are both useful, but the paper-knife (breadth half-an-inch) is by far the best for general purposes. Remember that though the animals have great powers of reproduction (it is said that we may procure two perfect anemones by cutting one across vertically or horizontally), yet, as a general rule, if they are at all lacerated they mortify, corrupt the water in which they are placed, and finally die. The net is used for lifting the animals from the pools, and for catching small crabs, shrimps, &c., which they will have much pleasure in devouring at a future time.

This zinc can (plate 11, fig. 5) is 8½ inches high, and its width is 8 inches by 4½. It is made of perforated zinc, in order to reduce the weight, no small object at the close of a long day's ramble: it contains one large, wide-mouthed glass-bottle, about 3 inches in diameter, next two small bottles about 3 inches high, and next to these two more bottles 5 inches in height—all wide-mouthed, well-corked; all the bottles are kept in their places by circlets of brass wire. The space above the four smaller bottles is filled by a zinc (solid) trough, to hold sea-weed, crabs and so forth. Any intelligent ironmonger will make a can of this sort, from the description and the engraving, at a very moderate cost.*

Of course a simpler and more economical plan would be to buy a glass or earthenware jar, fit a large cork, and place it in a small basket to reduce the chances of breakage, or even to invest in a small tin can (though this is liable to corrosion, and will not benefit your captives), but the advantage of carrying a number of bottles or jars is very great. There are many varieties of sea-anemones of a very small size and of delicate structure; these should have a receptacle for themselves, or they will be injured by the large heavy kinds, who, in addition to their own bulk, invest themselves with a coat, not of mail, but of stones and shells, which lacerates everything it touches—so, too, a *very* choice specimen can be carried home by himself, and the naturalist is not tormented by fears and apprehensions of what will become of his treasure in that solitary jar of "mixed pickles" which he dangles at his side. I have seen and heard of "vascula" made of wicker-work containing divers bottles and the requisite tools for extraction, but I consider them to be inferior to that which I have described, in consequence of their greater size and unwieldiness.

The rear-guard of our army is coming down the hill, he represents the corps of sappers and miners,

* The weight of the can and its contents when empty is three pounds and a half, and its cost about six shillings.

being an able-bodied man with a crowbar, and we will consequently send him on in the van when we commence operations. We can do nothing without our man and the crowbar; his office is to turn over all those large weed-covered angular rocks which lie at the verge of the ebb-tide—those stones which are never moved even by the roughest weather, and under whose sure protection lie all the rarest and most delicate anemones, besides all manner of other wonders which have not a place in our present records. To show you how great a difference so small a matter as a bar of iron may make, I will tell you the experience of two parties of naturalists who left Ilfracombe for Lundy in the summer of 1851. They went down in the same steamer, and searched the same rocks (Lamitor and Rat Island) at the same time. The first division despised crowbars, and thought their eyes were more to be trusted, but unfortunately, though they could probably see through a brick wall as well as their friends, they could not see through a sea-weed covered rock, and if they could have done so they could not get at what lay beneath; so their experience was, that they returned almost empty-handed, and preferred the Ilfracombe tunnels to the Lundy crags. Well, the second party were wiser, as it proved, and consequently exported a man and a crowbar, and by dint of diligent stone-turning for the space of two hours, they were able to return laden with all imaginable and unimaginable spoils. For instance, they procured numberless varieties of the "thick-skinned" anemone, him of the opal spots and the myriad-coloured arms; two varieties of "the gem;" two of "the daisy;" two of the beautiful "snowy anemone;" one of "the orange-disked," and two of "the orange-tentacled," anemone, neither of which varieties had been previously discovered on this coast; one of "troglodytes," of which more hereafter; and then there were divers animals of higher organization, which I will only name in the hope that some other naturalist will be tempted to explore the same prolific hunting-ground. Here they are, and I must premise that the creatures are far more beautiful than their names:—*Doris Johnstoni*, *Chiton marginatus*, *Pleurobranchus plumula*, *Botryllus violaceus*, *Ophiocoma rosula*, *Asterina gibbosa*, *Echinus sphæra*, *Pentactes pentacta*, *Terebella conchilega*, and many more. Now, almost the whole of these treasures were lying snugly concealed under big stones, which nothing but a bar of iron *could* have moved. So after this I hope you will not despise my man and his crowbar, although he is sitting at the edge of the Point, dangling his legs over the cliff, and smoking a very short and very black clay pipe.

Before we go down to the rocks let me say a few words about the tide, for its state is a very important item in the calculation of our probable success. The majority of the anemones which we intend to find are to be discovered at very low water—so low that we can only disturb their haunts once in every fortnight, and very many of them live at a still lower level. I do not mean to say that in ground which has not been previously searched we shall not meet with isolated specimens of most varieties at an ordinary state of low tide, but that if we mean to ensure success we ought to select those rocks which are only uncovered at the lowest tides. Now at new and full moon the "spring tides" occur, and then the sea rises higher and falls lower than at any other time, and at these periods it is "low water" on our western coast about twelve o'clock at noon; whilst, at the moon's second and fourth quarters, we meet with the "neap tides," and then the rise and fall of the sea is comparatively trifling. Therefore, if we intend to hold a grand field-day at any time we shall do well to select those three or four days which occur twice a month, when, according to the almanac, the moon is either new or full; and then, about twelve o'clock,* with the aid of our able-bodied

* If the naturalist be sufficiently enterprising to go down at twelve o'clock at *night* with a lantern to the rocks, he will be repaid for his exertion. On such an expedition two facts struck me as worth recording: one that *all* the "gems" were in full blossom; the other that every anemone, and especially the thick-skinned, were much more easily detached than in the day time: though in perfect flower and in full activity, they readily yielded to the slightest force. In about half-an-hour my collection exceeded that of an ordinary afternoon.

crowbar, we may reasonably expect that a tolerable harvest will crown our labours. The depth of the fall of water at spring tides varies considerably, but, by the aid of a Nautical Almanac, we shall readily arrive at a knowledge of the extent of shore which will be open for our exertions. On the shelving coast of North Devon a perpendicular fall of four-and-twenty feet will yield us as low a level as we can desire.

Very well; now, able-bodied man and crowbar, to the front—march! Don't stop to put out your pipe—and we will follow with the can and the hammer-case, and the whole paraphernalia of our shore-going warfare.

We descend this slippery path to our right, and arrive at a narrow ledge of rock, some two feet in width, with an uninterrupted descent on either side to the jagged rocks below. Never mind your nerves; forward! and a few roughly-hewn steps bring us down to the base of the cliffs. We will now trend to the westward, in order to hunt the extremity of that point which we have just left. Crash! splash! Oh! I see, you have fallen into a rock-pool, and are wet up to the knees, and have possibly sprained your ancle or dislocated your thumb. Yes, one cannot be too careful on these occasions, the sea-weed is extremely deceitful, and you ought to have "sprigs" in your boots. "Your ardour induced you to make a rush at the first sea-anemone you saw?" That was excusable perhaps, but not judicious. It certainly is an anemone, that soft, brown, jelly-like mass with a circle of turquoises round its mouth, and a blue stripe encompassing its base. It is *Actinia mesembryanthemum*, the "common" sea-anemone, and we shall find many varieties of it before we proceed much further. It sticks to the surface of smooth stones, sometimes in running water, and frequently where it will be high and dry till the tide returns: it is found in exposed situations at the highest level of all its brethren, and appears to rejoice in as great an amount of the "vital air" as it can manage to procure. Here is another variety of the same kind, a dull olive-green, with similar blue spots, which, however, are not very visible, being concealed by its expanded arms or tentacles. And here is "the strawberry," whose body is mottled with red and green, after the fashion of that pleasant fruit.

These three kinds are easily removed, without the assistance of chisel, hammer, or any other violent persuasion; we have simply to insert the ivory knife between their bases and the rock, and a very little exertion will deposit them in the depths of the largest of our glass bottles.

Another five minutes' scramble over rock and weed, and we come to the "half-tide" level, where, if we are fortunate, we ought to find two or three varieties of another species. Here, you see, the rocks are flat and interspersed with tracts of sand,

and there is what the Devonshire people call "a fleet," that is an incline down which the water trickles gently to its level. Further on a barrier of rock and shingle forms a large and shallow pool.

It is very beautiful, perfectly clear and transparent, mirroring every cloud-shadow, and reflecting the glare of the sun, so that at first we can see little but its wind-ruffled surface. The dense, pink-hued coralline-tufts line its margin, then the delicately-lobed, waving foliage of the green laver, and beyond the dark crisp thickets of the Carrageen moss, whose every branch hurls back a changeful, many-tinted rainbow of light. Here at the verge of the pool we find a noble prey, "the waxen-armed Anthea," *Anthea cereus*, or, as he is unpolitely surnamed in these parts, "Legs:" his body is of an olive-green hue, and thence depend a forest of long, waving, snake-like green tentacles, with bright purple tips. Next to him is another variety, slate-coloured throughout, and some day we may be fortunate enough to meet with a specimen which shall be white as the fallen snow. This *Anthea* is generally found in shallow, sunny pools, and quite close to the water's edge, so that he too is fond of air, though not as much so as his neighbour, the "common" anemone.

Still lower, here under the shadow of an over-hanging rock, is a sheltered corner, and a bed of fine shingle just covered with water, and a sight which

W. Brodrick, del. London, John Van Voorst, 1 Paternoster Row, 1856. G. H. Ford, Chromolith. W. West, Imp.

Actinia Coriacea

you will not forget for many a day. What a magnificent "bloom!" it is too large for a "show" Chrysanthemum, and too gorgeously arrayed; it is a good four inches across, pearly-white in the centre, then a broad ring of translucent lake, and then the petals!—row after row of transparent tentacles ringed with lake and delicate brown, and pearly white. Does not this come up to your "heart's desire?" If not, you must be very hard to please. Ah! you have touched it, and it has vanished entirely; we can see nothing but gravel, and a little crab, which is scuttling away sideways, in a state of pugnacious amazement. Now, if you put your hand down upon the place where you last saw the flower, you will feel a round lump of gravel; scrape away all the loose stones round it, and then with the knife detach it very gently from its resting-place, or, if the rock be soft, chisel it away as patiently as you can. This is *Actinia coriacea* (the "thick-skinned" anemone), and he puts on this coat of stones to conceal and defend himself, and every stone is attached by means of minute suckers, with which his body is covered. Recollect that the least injury done to this anemone is generally fatal, and also remember that you may frequently discover him when he is not expanded, by *feeling* diligently in shingly nooks and corners just covered by the water, or even at the bottom of deep pools. There are innumerable varieties of this kind; the body is generally marbled with red and orange, and always studded with large white or opal tubercles or warts, but the colour of the disk and tentacles—*i. e.* of the expanded bloom, —is infinite in its changes: russet-green, crimson and white, red and orange, heather-tinted, dove-colour and lake,—every possible gradation of these and every other colour, though I can't say I ever saw a *blue* sea-anemone, which is as great a desideratum as a blue dahlia.

Having arrived at the extremity of the point, and at the lowest level of the lowest tide, we will set our crowbar to work. Here is a large rock, well sheltered from the Atlantic swell, well covered with coralline and sea-weed, and well situated in the midst of a large pool, so that its base is always under water in all weathers. Now, heave and with a will! Over it rolls! Away rush half-a-dozen bull-heads, and a young conger, viciously writhing, and lashing his tail as though he would do us a mischief if he could; and only look at the crabs! Let us pick up that flat, hairy, dingy little rascal: he is a curiosity, by name *platycheles*, or "the broad-clawed," and very broad his claws are, for the pair of them side by side would more than cover his whole body: he is of a rather sulky disposition,—a little crabbed, we should say, if it wasn't for the apparent pun,—but he will lie harmlessly enough with our anemones when we get him home. Here is a pink button sticking to the stone, streaked with vertical lines of small opal dots or tubercles, six or more of these lines being very distinctly composed of larger tubercles of an opaque white colour. This is *Actinia gemmacea*, "the gem," or, in the vernacular, "buttons." He will disclose about fifty snaky, barred and spotted tentacles, and his mouth is of a brilliant green, and there are half-a-dozen other colours you will discover in him when he is well opened to-morrow. This is a treasure; it is common enough on the North Devon coast, but almost peculiar to that locality.

We turn over half-a-dozen other stones, with perhaps little success, so far as anemones are concerned, but we bag a few star-fish and a few compound zoophytes. By the way, people are always saying that every anemone is a zoophyte, which is true, and consequently inferring that every zoophyte is an anemone, which is not true. Every jackdaw is a bird,—true; then every bird is a jackdaw,—similarly not true. We have seen that the term "polyp or true zoophyte" * includes a vast class of animals,—Hydras, Sertularias, sea-fans, corals and sea-anemones, — therefore every anemone is a zoophyte, but every zoophyte is not an anemone.

Excuse this digression, and over with another stone. This one, you see, is full of holes, bored by a little mollusc called the *Pholas;* and there is a little pink wart in one of the holes—chisel him out

* Some "Zoophytes" (*e. g.* ascidian) are referable to the Molluscan kingdom.—Johnston.

—he is *Actinia troglodytes*, or "the cave-dweller," so named from a race of African shepherds who lived, like our friend, in caves hollowed by other hands. He is a rarity on the North Devon coast, having been first discovered near Ilfracombe by the author in the summer of 1855, and comparatively few specimens have as yet been found, but he is common at Tenby, in "the caves," where he frequents the company of *A. nivea* and his allies (Gosse). Other habitats are Berwick Bay (Johnston), Cornwall (Couch), Isle of Man (Forbes), Moray Firth (Robertson).

We wend our way further to the westward, and halt under a shelf of rock, to examine three or four narrow pools which lie far back beneath the overhanging roof, on a horizontal section of the slate. Here we find "the daisy" (*Actinia bellis*), a dark gray anemone, with numerous streaked and mottled tentacles. It dwells in angular cavities, and shrinks far out of sight on the slightest disturbance. It is known chiefly by its flat gray disc and its cup-like body. Chisel it out, then, and put it into one of the smaller bottles. There is a variety of a rich chocolate colour, with an underlying tint of dull crimson. This species generally inhabits shallow, sheltered rock-pools, some few feet above low-water mark.

See! the tide has turned, and the sea-fog is stalking in slowly from the Atlantic. We have done a capital day's work, for which we may thank, first, our crowbar, next, the ground-swell of yesterday, which rolled in many of our choicest specimens, and lastly, the fact of our having searched for each species at its proper level and in the peculiar situations which it loves to inhabit. Now we will add a little fresh and clear water to each of our bottles, and chisel off a few pieces of sea-weed, this plant of green laver and a few fronds of Carrageen moss, and start homewards, resting for awhile under the broken cliffs of Arragonite Bay, to recruit our strength and see that our specimens are completely settled in their temporary abode.

I am not going to moralize. See how the gulls are soaring in wide circles overhead, screaming in the fulness of their delight, satiated as they are with the dainty repast of the ill-starred limpets and the mussels which they have devoured at low-water mark; and what a strange wild laugh that is of theirs! how often one hears it at early dawn when they are flying far overhead in misty flocks to seek their inland pasture-grounds. There is a cormorant, with outstretched neck, working his way up channel in straight and measured flight: his shadow flits by us, a dim gray cross on the gray stones. And there is a windhover poising himself motionlessly with outstretched tail and curved wings, and his head well up in the wind: down he comes with a swoop, and now he is again in his old station, as though the intangible air were a quiet resting-place. I am not going to moralize, and so I shall not make any remarks about the wonderful, and perhaps, unexpected beauty which we have discovered in these lower forms of creation. If a man cannot discover for himself the evident beauty and design and knowledge which exist in every rock-pool and on every weed-covered stone, no amount of talking or writing will drive the palpable fact into his head; and if, after such an excursion as ours of to-day, he do not return home a wiser and a better man, he must lay the fault where only it is due—at his own door. But I must add, as we stroll homewards, that one great benefit to be derived from the pursuit of Natural History at the sea-side, is the intense relief and the renewed buoyancy which it grants to a mind wearied and overtasked by the realities of daily life. Do you remember how mournfully Keats sings of

> " The weariness, the fever, and the fret
> Here where men sit and hear each other groan;
> Where palsy shakes a few sad, last, gray hairs;
> Where youth grows pale, and spectre-thin, and dies;
> Where but to think is to be full of sorrow"?

Poor Keats! I wish I could have taken him out anemone-hunting after these sad lines had struggled out of the recesses of his heart. There is no "companion of our solitude" like the sea:

> " There is a rapture on a lonely shore,
> There is society, where none intrudes,
> By the deep sea, and music in its roar."

There is no state of mind with which it cannot sympathise—no trouble or anxiety which it will not comfort and allay. So eternally calm—so gloriously fresh and vigorous—so pure, and free, and infinite! And these silver-gray, sun-lit rocks, and this heaving, glistening forest of sea-weed, and this shining tract of white sand and many-tinted shells—how full of perfect beauty, how full of endless life!

Sad heart, wearied with the petty jealousies, and envyings, and bickerings of thy fellows—maddened by the narrow-minded strife of empty, meaningless words—sated with the daily slaughter of truth under the shallow pretence of zeal for art, and morality, and religion—leave for awhile the reckless whirl and the bloody arena of the world of men, and come and learn a purer philosophy from the mute eloquence of God's lower creation; learn a calm simple trust in to-morrow's sun and to-morrow's sustenance—a quiet fulfilment of the daily routine of daily work—an unvarying love for a life of faith, and gladness, and adoration.

CHAPTER III.

WHAT IS ITS NAME?

TIME: *after breakfast.* SCENE: *if the drawing-room, enlivened by the mingled curiosity and disgust of the lady portion of the family, who do not approve of the unlimited introduction of crocks and salt-water among their crochet and water-colours, and are apprehensive lest the creatures should walk about the room and bite: if the study, so much the quieter.*

Last night we turned out all our treasures in two broad pans with plenty of sea-water, the larger specimens in one, the smaller in the other. To-day, then, I wish to show you, as simply as possible, the easiest way * of finding out the various names of the captures which we have made or shall make at some future time.

* In this Chapter the subject is treated as popularly as possible: in Chapter VI. the reader may see a statement of some of the numerous difficulties with which it is invested.

We fill three shallow pans or glasses with fresh sea-water, and proceed to sort out our specimens.

In No. 1 we put all the anemones which have smooth, soft skins; and we put them in with more certainty if we find they have a circle of beads outside their tentacles and a blue line round their bases.

In No. 2 we place all those specimens which are covered with large, conical, coloured tubercles or warts.

In No. 3 we insert all the rest, observing that they have less prominent warts, which are disposed only upon the upper half of their bodies, and that if we stir them up with a paper-knife they will probably shoot out a number of white threads from all parts of their body, meaning thereby a personal remark to us about our unnecessary intrusion.

All the *Antheas*, or "Legs," we leave in this receptacle, for they are not Anemones, because?—

Because they have not the habit of drawing up their tentacles within their bodies when danger threatens them.

Whilst we are engaged in sorting out the specimens, I will describe briefly the construction or the anatomy of the sea-anemone, for when we come to define the various species we shall have to speak of "oral discs," "thread capsules," and the like, and it will be well to begin by understanding what these terms mean.

Take a tumbler and place a circle of card inside it, just below the edge, cut a round hole in the centre of the card, and stick two or three circles of pins, heads uppermost, on the card's upper surface and as close to the glass as you like. Here is a model of an Anemone, rough certainly, but sufficiently accurate for our purpose.

That part of the tumbler which touches the table corresponds with *the base* of the live animal, the outside of the glass with its *body*, and the edge of the tumbler therefore with the *margin* of the body. The whole of the upper surface of the card represents the *oral disc* [" disc," a flat circle; "oral," round the mouth].

The hole in the centre is *the mouth;* draw a line round it, and it will include *the lips.* The circles of pins are the circles of *tentacles,* and their heads represent the *tentacle heads.*

That is clear enough: now for the inside of the animal. In Plate II., fig. 4, is a section of a full-grown specimen of the "thick-skinned" anemone. At *b* is the outer skin, stretching from one margin of the body, round the base, to the other margin: *a a* are the tentacles, situated on the oral disc which terminates at *d*, the mouth, and below *d* is the *stomach,* which some of your specimens are sure to protrude for your inspection at some time or other, so I need not describe that.

The so-called "*ovaries*" lie coiled round the

stomach (at *e*) in separate chambers,* which are represented in another section at *g*. These chambers communicate with one another by openings (as at *f*), and with the tentacles, as shown at *c*.

All this we can see with our naked eyes, especially as the large anemones frequently swell themselves out with water till they become perfectly transparent. But there is one small piece of microscopic observation which it is quite necessary to record. In one part or other of the body of *every* sea-anemone may be seen a number of pointed, oval cells (shown at *a*, Plate II., fig. 1), containing a thread or spike, which can be hurled out from one extremity (as at *b*), but is usually folded back within its case. These cells, which I shall call "*spike-cases*" (Mr. Gosse speaks of them as "thread-capsules"), are found within the blue beads of the common anemone, on the lining-membrane of the ovaries in the "thick-skinned" and the "gem," and within certain white threads which the "daisy" and the "snowy" anemone send out

* Since this was in type, my friend Mr. G. H. Lewes has added, if possible, to his scientific reputation by the important discovery of the *real* ovaries of the *Actiniæ*. A short time ago he dissected an anemone in my presence, and pointed out the ovaries lying in the interceptal spaces, and *completely hidden by the convoluted bands*. After which, to remove all doubt, he placed a portion of the ovary under the microscope, and showed me the unmistakeable ova lying in it. Now, in the so-called ovaries,—that is, in the convoluted bands,—no ova are to be found, whilst spike-cases are abundant, and these latter are entirely wanting in the real ovaries. This discovery leaves the function of the convoluted bands a problem to be solved.

from every part of their surface, and are abundant in the tentacles of all. Observe that each of our three pans holds a group of anemones, which carries its spike-cases in one position peculiar to itself—the beads — the ovaries — the white threads. See Chapter VI.

A few words about these white threads of the daisy and his friends. If you incommode "daisy" with your finger-nail, he will shoot out his threads at you, as I have said; cut off the end of one of them, and put it under your microscope, and it will present the appearance represented at *a a*, Plate II., fig. 2, a round white cord, moving freely by means of multitudes of invisible "cilia," or hairs, with which it is fringed, and composed of an outside membrane, to which the cilia are attached, and which invests innumerable quantities of the round cells shown at *c*, fig. 1, and great numbers of the "spike-cases" at *a* and *b*. When the anemone recovers his temper he draws back his threads into his body. About the *use* of these "spike-cases" I can give you but little information. Naturalists believe them to be weapons of offence, and that the spike is exercised in killing the animalcules, crabs and molluscs on which the anemone feeds, as well as in warning off any hostile intruders. See Mr. Gosse's 'Devonshire Coast,' for the observations on which he grounds this opinion.

Now for our specimens in pan or glass No. 1.

First, they all have smooth soft skins, and if we

use the paper-knife skilfully we shall see a circle of blue beads inside the margin of the body, and therefore just outside the outer row of tentacles. When the animal is in full flower, the beads are generally visible. All anemones which have this peculiarity may be placed in a group by themselves (if you require a *name* for the group, see Chapter VI.), and there are three species, two of which are common on our coasts.

The first is *Actinia mesembryanthemum*, "the common sea-anemone." Its body varies considerably in colour: we may obtain varieties ranging from light to dark green—some are brown, some liver-colour (and this is the ordinary variety), some light red, scarlet and orange-buff. Two varieties are striped, one an apple-green with pea-green lines, another scarlet banded with yellow. The colour of the beads varies from dark to light blue, in one variety is white, and in young specimens a purple-gray. A blue rim round the base is generally clearly marked.

The "strawberry anemone" is usually supposed to be a variety of the common species, but, as may be seen in Chapter VI., I am inclined to think that it is an independent species. It is of a brown colour, with green spots, and the blue line of the base of the common kind is almost, if not entirely, wanting in the "strawberry." The most important difference is this, that the largest common anemone never approaches the size of the largest "strawberry." Now if one were a variety of the other, and both were living in like situations and fed in the same manner, it is at least probable that there would be no permanent difference in the point of size.

In our pan No. 2 we must remember that all our specimens are studded throughout the whole surface of their bodies (though more scantily towards the base) with large, conical, coloured, perforated tubercles or warts. We take up one of these animals, and ask, Are its warts arranged in vertical lines, and do six or more of these lines contain larger and whiter warts than the rest? If so, we may be tolerably sure that we have the beautiful *Actinia gemmacea*, "the gem." When closed its body resembles in shape an oval waistcoat-button. When open it discloses about fifty tentacles, long, large, of an olive-colour, barred on the side next the mouth by streaks of white. The colours of the oral disc are commonly gray on the outside, then yellow, and next bright green, which tint surrounds the mouth. The shape of the body when expanded is very variable, generally it is columnar, and frequently it rises in this shape to the height of one or more inches. The average diameter of its body, when in contraction, is half an inch.

Next, we will look in the same pan for a specimen whose warts are not arranged in regular vertical lines, and perhaps discover half-a-dozen or more which answer to the description. All these, then, are individuals of the species *Actinia coriacea*, the "thick-skinned anemone."* The body is generally marbled with red and green, colours which, in captivity, turn to a uniform dull red. The tentacles seem to be arranged in four rows, the first and second rows containing ten tentacles in each, and the total number being one hundred, more or less. The variety of colour is too great to be detailed within the limits of the present work, and I can only give two or three kinds as specimens of the whole.

1. *The Russet Thick-skinned.*

Oral disc dove-colour tinted with lake, shading into russet among the tentacles. Tentacles dove-colour tinged with russet, barred with white and Indian red. Lines of the two latter colours encircle the bases of the tentacles in the inner rows.

* Johnston, in his 'British Zoophytes,' recognises two species which are closely similar, *A. coriacea* and *A. crassicornis.* He says that the latter inhabits deep water, and never indues itself with a coat of stones. Mr. Gosse denies that this remark is ever verified in any species of the genus, and includes both kinds under the name of *crassicornis*, or "thick-horned." I have ventured to preserve Professor Johnston's original nomenclature, and have called the species which is so abundant on our coasts *A. coriacea*, "thick-skinned," certainly a more characteristic name than "thick-horned." If there be a "thick-horned" species which does not invest itself with stones, it has, at any rate, never been heard of in late years.

F

2. *The French-white Thick-skinned.*

Disc white, streaked with red and orange. Tentacles dove-colour, ringed with white and red.

3. *White-tentacled Thick-skinned.*

Disc crimson-lake, shaded into purple. Tentacles white, with rings of bright crimson surrounding those of the first row.

The naturalist may amuse himself by adding endlessly to this list.

The average diameter of the contracted body of this species is 1½ inch, and of its expanded bloom from 3 to 4 inches.

In our third pan, which contains the last group, we find that all the specimens agree in having minute flattened perforated warts on the upper surface of their bodies, and that they send out from invisible pores their ciliary white "threads" containing "spike-cases."

The first specimen which we examine has a cup-shaped body, that is, the disc is two or three times larger in diameter than the stem, when the latter is at all elongated. Generally, however, this species may be recognised by the extreme flatness of the disc, and by the vast number of small tentacles which are crowded together, and lie more or less horizontally upon the disc's outer edge. The body is of a flesh-colour, terminating in a dark gray upon the upper margin, and is spotted with gray warts of the kind above described. Disc gray, streaked and spotted with white. Tentacles (about 700 in number) the same, resembling, when under a lens, the spotted skin of a snake.

This, then, is *Actinia bellis*, "the daisy."

One variety seems to have a yellow horn projecting from the interior of the disc—in reality, one of the tentacles in the inner row is enlarged and coloured.

Another variety is of a rich chocolate tint, streaked with dark crimson.

Another has a disc of alabaster underlined with pure vermilion.

Next we come to *Actinia troglodytes*, "the cave-dweller." His body is usually of a pale flesh-colour, and sometimes of an olive-green, and towards the base is streaked with vertical lines of white, one broad band alternating with two or three narrower lines. His disc is of a grayish pink. The tentacles are about fifty in number, generally more or less erect, transparent, white, ringed with opaque white, and the base of each surrounded with a very dark brown ring, containing a heart-shaped spot of white.

We have thus named the majority of the specimens which we should probably find in the locality of which we have been speaking. We have also learnt a method of dividing the sea-anemones into three natural groups, or genera, as they are called, by which means our difficulties in the way of naming our collected specimens are considerably diminished.

I shall now conclude this chapter with a tabular arrangement of these three groups, and of some of the species which they contain, and which are usually to be found at one or other of our best-known watering-places.

Should the reader require a more complete and scientific arrangement he is referred to the Table which ends Chapter VI.

Group I.

Skin smooth; a row of beads on the oral disc, close to the margin of the body, which contain "spike-cases."

1. *Actinia mesembryanthemum*,
"The Common Sea-Anemone."

Skin soft.*

Varieties. *a.* Brown.
b. Olive-green.
c. Striped with pea-green on an apple-green ground.

* A soft, smooth skin, distinguishing it from the leathery, smooth skin of *A. margaritifera.*

Varieties. *d.* Leek-green.
e. Liver-colour.
f. Light red.
g. The Tiger. Yellow stripes on a red ground.
h. Orange-buff.
i. "Chiococca." Beads white; body scarlet.

2. *Actinia fragacea,*
"The Strawberry Anemone."

Skin soft, spotted with green on a liver-colour ground.

3. *Actinia margaritifera,*
"The Pearly Anemone."

Skin leathery, beads ultra-marine.

GROUP II.

Skin covered with numerous conical, large, coloured, perforated warts.

1. *Actinia gemmacea,*
"The Gem."

Warts arranged in vertical rows, six or more of which contain larger and whiter warts than the rest.

a. The common variety.

b. The glaucous variety (*A. thallia,* Gosse).

Disc, a many-rayed star of yellow on a blackish ground.

2. *Actinia clavata,*
" The Weymouth Anemone."

Warts arranged in vertical rows, and all equal.

a. The yellow }
b. The rose } variety.
c. The glaucous }

3. *Actinia coriacea,*
" The Thick-skinned Anemone."

Warts irregularly arranged. Varieties numberless.

Group III.

Warts flattened, and on the upper portion of the body only. White threads sent out from invisible pores, and containing spike-cases.

1. *Actinia bellis,*
" The Daisy."

Disc flat, two or three times as broad as the body when the latter is at all lengthened. Tentacles very numerous, small and crowded horizontally on the outer edge of the disc.

a. Common variety.
b. Yellow-horned variety.
c. Disc ivory-white.
d. Disc chocolate and crimson.
e. Disc alaabster and vermilion.
f. Disc rose-colour.
g. Disc and tentacles amber-colour.

2. *Actinia troglodytes,*
"The Cave-dweller."

Tentacles erect, few, taper, barred at their base with a dark ring, containing a white heart-shaped mark. Habit, hides under stones; body irregularly flattened when in a state of contraction.

3. *Actinia anguicoma,* or *viduata,*
"The Snake-locked Anemone."

Tentacles few, very long, taper and flexible. Two vertical streaks* down each tentacle. Habit, alternately very flat and columnar.

4. *Actinia rosea,*
"The Rosy Anemone."

Tentacles few, rosy red; mouth a cross of four rounded lobes.

* These vertical streaks are found in *A. troglodytes.* I am inclined to believe that the two species are identical. See Chapter VI.

5. *Actinia candida*,
"The White Anemone."

Tentacles graduated, those of the inner row being the largest, opaque snowy white. Inner row of tentacles ringed with dark red or brown. Body and disc white.

a. Common variety.
b. *immaculata*. Wholly white, the red ring round the inner tentacles being obliterated.

6. *Actinia dianthus*,
"The Plumose Anemone."

Margin of the disc fringed with numerous small tentacles and much lobed. Tentacles many, graduated. Body from 1 to 5 inches in diameter.

a. White variety.
b. Cream-coloured.
c. Flesh-coloured.
d. Pale red.
e. Orange.
f. Olive.

7. *Actinia parasitica*,
"The Parasitic Anemone."

Outer tentacles smaller and turned downwards over the rim of the disc. Tentacles about 500 and graduated. Body about 2 inches in diameter and

3 in height, of a yellowish white, ribbed vertically with dark brown. Habit, parasitic, chiefly on shells.

8. *Actinia Aurora,*
"The Orange-tentacled Anemone."

Outer tentacles smaller, and turned downwards. Tentacles about 80 in number, tinged with orange.

Body ½-inch in diameter, of brown or olive-green colour, lined with vertical bands of white.

9. *Actinia venusta,*
"The Orange-disced Anemone."

Outer row of tentacles smaller than the rest, and turned downwards. Tentacles about 250, graduated, white.

Body ½-inch diameter, orange-brown. Disc oval, of a bright orange colour. Base oval.

a. Common variety.

b. *nivea,* "the snowy anemone." Disc French white; body orange.

Here, then, we conclude our enquiries into the "names" of our friends the Anemones, and I trust that, if the amateur-naturalist uses this table in conjunction with the more lengthened remarks which precede it, and the observations on the respective abodes of his specimens in Chapter II., he will not find an insuperable difficulty in identifying the majority of his collected treasures.

The varieties, and I may add the species, run one into another, equally perhaps with that difficult tribe of plants, "The Ferns and their Allies," and therefore nothing short of lengthened and accurate personal observation will reduce this branch of the subject to a certainty. There is no greater difficulty in Natural History than the determination of specific characters; and if the student be not always satisfied with the result of his own and others' labours, he will, at any rate, have the consolation of exercising on his road three great and practical virtues—

PATIENCE, ACCURACY AND PERSEVERANCE.

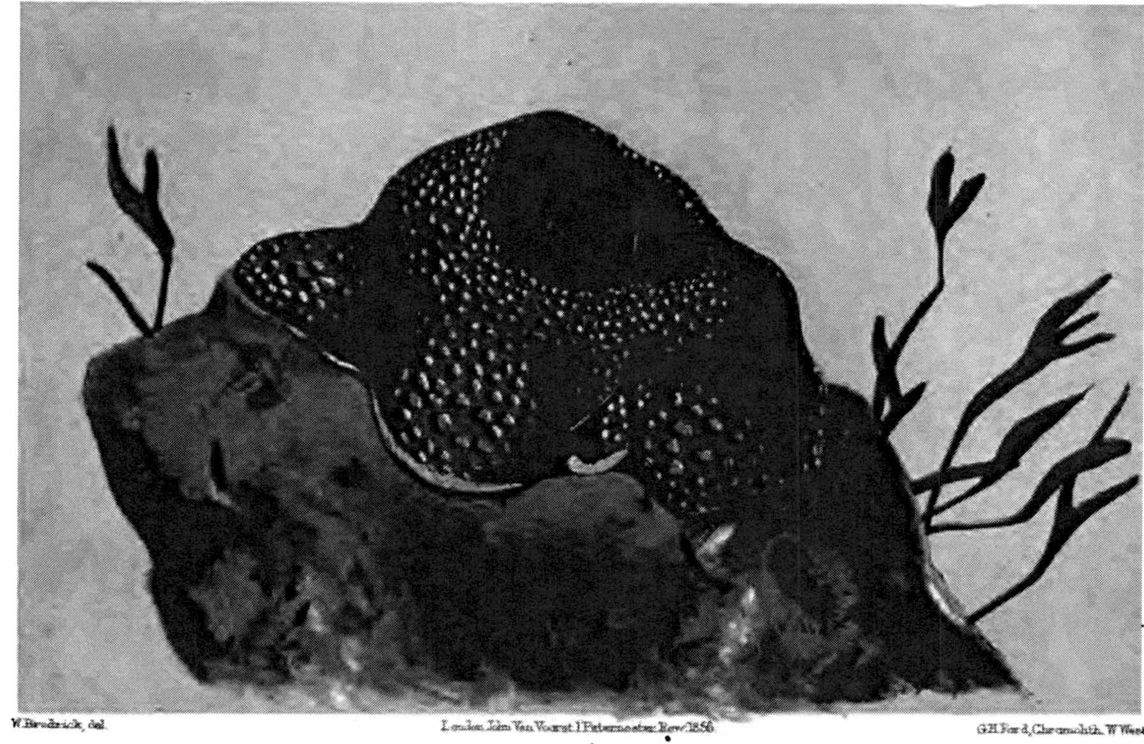

W. Brodrick, del. London, John Van Voorst, 1 Paternoster Row, 1856. G. H. Ford, Chromolith. W. West, imp.

CHAPTER IV.

HOW SHALL I KEEP IT ALIVE?

WERE you ever led, reader, by chance or by choice, into one of the plague-courts of London? I do not speak of the Black Death of the fourteenth century, but of that pestilence which is hardly less fatal in our own times, the plague of neglected poverty,—starving on mouldy crusts and fiery gin,—choking in a poisoned atmosphere,—wallowing in the accumulated filth of countless years.

Have you ever trodden those crowded, mouldering lanes and alleys, where open sewers—witches' cauldrons of festering filth—seethe and welter by the open doors,—nay, roll their rank pollution through the very heart of the poor man's home; where vermin, unnamed and unknown in civilized life, creep and writhe, and die and rot, on wall and floor and roof—a moving, mortifying crust of life and death—the mockery and bathos of the decorative art; where the sickly glare and the wearied smile of consumption ape the glance and the laughter of health; where the strong grow weak, and the weakly bow the head and die; where the innocence of the child is taught to curse, and lie, and steal; where the pride of manhood is quenched in the imbecile leer of the sot; where the fair honour of womanhood is sullied like the snow which falls in those infernal regions; where God is as unknown as the pure air of His own heaven?

I do not pen these words for writing's sake, and I have not exaggerated a fact to point an antithesis or construct a climax. You may find all and more than this verified in the records of the past summer of 1855; but I speak of things as they have been, and are, and should not be, and I lay them now before you in order to draw out the causes whence so much misery and crime originate.

Man, taking him in the lowest point of view, is an animal. But he is an animal of a high organization — his internal machinery is more adapted than that of any other animal to elaborate great results: the fact is, he has a mind and a soul, and his body is constructed accordingly, so that these diviner portions may be placed in full relation with external objects. He possesses a complicated breathing-apparatus, which we call his lungs, and these, in their normal condition, inhale the oxygen of the air to purify his blood, and give out the carbonic acid gas, which is a portion of the *débris* of his structure: he has a series of organs which receive, separate, and apply certain organic substances, which we call his food, to build up his frame and supply the waste which is caused by his daily muscular exertion, or his daily work; he is blessed with a nervous system which enables him to communicate with the external world, which receives impressions, and originates motion, and through which he thinks and wills.

Very well: we place this delicate piece of mechanism in a damp, plague-stricken abode, laden with *in*-organic impurities of all kinds,—where his lungs cannot imbibe the amount of oxygen which they require,—we feed him with substances from which his digestive organs cannot derive sufficient materials to increase or support his frame,—we insist upon surrounding him with objects which can afford him no healthy sensation,—we then have the assurance to marvel that, having breathed, and fed upon, and felt the essence of filth for a number of years, he can have the audacity to think and act filth for the remainder of his natural life. The marvel would be were it otherwise. We should not, in our sane moments, take the case off a clock, set it in a dusty highway, and expect it to strike the hours with its wonted regularity, nor, in like manner, can we expect a man to work as a man when we abstract from his life the very conditions of his manhood.

Now, if this be true of a man, it is, to a lesser extent, equally true of all animals, and therefore of a sea-anemone. If we desire to keep any living object in a state of health we have only to surround it with those conditions of existence in which it was placed at the creation by its Maker, and our object will be accomplished, so far as any foresight of ours can effect our purpose.

I make these prefatory remarks, because I do not wish to be considered a quack-doctor of anemones. A quack is a person who lays down a set of rules on the subject of health, or perhaps on any other subject, which rules he makes a point of not explaining: he appeals then to a man's faith and not to his reason, on a topic where faith is doubtlessly advantageous, but in which reason is capable of being exercised. Hence he places himself in a dilemma: either he is not himself aware of the reason of his practice,—in which case his knowledge is exhausted together with his rules,—or he is acquainted with first principles whence his rules originate, in which case he insults the common sense of the world by refusing to allow that it can grasp his arguments and appreciate his experience.

It would be sufficiently easy to string a series of rules, which being duly observed would enable a man to keep himself, his fellow-man, his dog, or his sea-anemone, in a state of health for an indefinite period under ordinary circumstances, but at the same time, for lack of understanding the reason of his proceedings, he would be liable to omit perhaps the most important condition of health, or, should unforeseen accidents arise, would be totally unable to meet the emergency.

With regard to our sea-anemones, the "common sense," then, of the matter is simply this: they, being animals, take in large quantities of the oxygen which exists both in the water and in the air. This gas is as necessary to their existence as to ours, and when they are deprived of it they languish and die, as we do in close rooms and pestilential courts. They feed on the animalcules which abound in the sea-water, consequently, if we continually *filter* the water, although we may purify it from all decaying and refuse matter, we deprive our captives of a large amount of the food which is also necessary to them.* The larger anemones (especially the thick-skinned) seem to require more sustenance than the minute animalcules can afford them, and in their native depths they devour crabs, shell-fish and the like,

* The question then arises, how do they flourish in Mr. Gosse's manufactured sea-water? The fact is indubitable; how do we account for it? Mr. Gosse puts in his weed some days before he puts in his animals. Now, if we place some fresh water in a clean glass vessel, and throw in a few clusters of pond-weed or the like, in a few days the microscope will disclose countless numbers of animalcules which did not exist before. Doubtless then, in both cases, they spring from the invisible eggs or spores which abound in the atmosphere and the water, and which vivify under the proper conditions of existence. So mites are born in mouldy cheese, "eels" in paste, vinegar, and so forth.

consequently we must indulge them with such delicacies whilst in captivity, or they will gradually fall into an atrophy, dwindle away almost imperceptibly, and after the lapse of many months, though they be apparently in perfect health and in full bloom, will have diminished to a half or a quarter of that size in which they rejoiced when we chiselled them out of their rock-pools.

All the anemones throw off daily a coat of slime, and this is chiefly observable in the case of the common species; they also cast away the refuse of their food in small pellets, which are ejected from their mouths, linger among the tentacles, and fall to the base of the rock on which they adhere. If we keep sea-weed amongst them, this too decays and forms a slimy deposit. The minute Serpulæ, Nereids, and other animals which inhabit the stones we place in our aquariums, die speedily and corrupt the water. Now the Marine Board of Health disposes of all these nuisances in an instant: up rushes a large wave or a heavy under-current, and sweeps away every particle of decaying matter, dashing it on the rocks again and again, until, by constant contact with the air, the whole mass of impurity is thoroughly purified.

Every wave which breaks upon the beach, or curls its foam-flecked crest in multitudinous light-spangles far out at sea, contributes somewhat to this great sanatory work, and adds its offering of vital air to the mighty ocean world.

Another point of importance is temperature. It is well known that the extent of the variations of temperature is less in the water than in the atmosphere. In the temperate zones (lat. 50° in the part of the Atlantic which is nearest Europe), whilst a thermometer in the latter medium ranges from 66° to 35°, in the former the variation extends only from 68° to 41°. Therefore, in imitating this natural law, we must guard our captives from lengthened exposure to the sun in the summer time, and in winter from very hot rooms, and from the extreme cold of the external air. An average temperature of 55° Fahrenheit will agree very well with our friends' constitutions.

The last caution I need give is with regard to light. Water is less transparent than air: it is denser, and transmits light with more difficulty. All the anemones habitually live during a certain portion of the day in a state of comparative twilight But many of them (*e.g.* the "daisy," the "cave-dweller," the "gem") conceal themselves in nooks and angles of the rock, and beneath huge stones, where little if any light can affect them. We must then endeavour to imitate this condition of their existence by the aid of a north aspect, or of shaded receptacles containing abundance of angular rocks or corals.

Let me recapitulate the five points of our treaty with the anemones:—

First. A plentiful supply of oxygen, which is contained most abundantly in pure sea-water, but sea-water is purified by admixture with the oxygen of the atmosphere, therefore we shall have to give them *a plentiful supply of pure air.*

Secondly. *Food:* either the animalcules contained in water just taken from the sea, or those which are generated in pure and unfiltered salt-water, and, for the larger animals, shell-fish and the like.

Thirdly. The *removal of all decaying matter,* whether vegetable or animal. By this I mean the manual removal of those impurities which are not corrected by the addition of sea-water, or by aerating that which is already in our receptacles.

Fourthly. An *equable temperature.*

Fifthly. *Moderate light.*

If we ensure these five conditions to the anemones all other considerations appear to be matters of the smallest importance to them: they are totally indifferent whether we keep them on the Devonshire coast or in the heart of London—whether we give them fresh sea-water daily or aerate that which we gave them a year ago, or immerse them in a saline fluid manufactured at the nearest druggist's shop: they are quite regardless of the fact of their being admired by lords and ladies in a palatial aquarium, and flourish with equal vigour and beauty in the earthenware pan and the white-washed cottage of the less aristocratic Nature-student.

But, lest in my desire to avoid the Charybdis of the meaningless rule-systems of quackery, I fall into the Scylla of vagueness and inaccuracy, I will add a few directions which shall enable the reader to carry out these life-principles with as little expense and trouble as may be possible.

First, let me speak of the *receptacle* in which we intend the animals to flourish. For general purposes there is nothing like glass, and among vessels of all shapes and sizes the best is the ordinary finger-glass of our dinner-tables; a few pebbles in the bottom of it, and a light paper cap to shield it at times from light and dust, will render it ready for use: tumblers of thin common glass are good for small specimens: confectioners' show-glasses are not to be despised, especially when we desire to contrast the long green fronds of the *Ulva* with the vivid colours of our sea-flowers. My *beau ideal* of a receptacle to stand on the naturalist's study-table is a glass (of the thinnest and whitest material), 9 inches in diameter, 4 in depth, with a perfectly flat bottom, and standing on a foot-stalk of 3 inches in height, to admit of the working of a small reflector beneath. A square glass trough of similar dimensions might be in some respects superior, though not so durable. It may be constructed of slips of plate-glass, cemented together with "marine" * or "liquid" glue.

* To make marine glue, dissolve caoutchouc in oil of coal-tar, and add two parts by weight of shellac for each part of the solution. The composition is to be used at a heat of 228° Fahr.

Earthenware vessels, brown pans, pie-dishes, foot-pans, are useful for the commoner sorts of anemones. A stone-trough, 2 feet by 14 inches, makes a capital preserve or nursery, when placed under the shelter of an alcove or summer-house, and the more porous the stone is the more equable (during the heat of summer) will be the temperature of the water it contains. It should be set on a slight incline, and have a hole bored at one extremity with a cork fitted to it, in order to drain off the water for cleansing purposes. I will briefly describe a more elaborate receptacle of this latter kind, and then leave my reader to exercise his ingenuity in devising "aquaria" for himself.

Take a large slab of oolite, say 4 feet by 2, and 2 inches thick; bore two holes in it, one ½-inch diameter in the centre, the other an inch in diameter close to one end, and fitted with a cork, or a pipe and tap. Let this slab form the bottom of the tank, groove it, therefore, and fix upon it, with Portland cement, four sides of stone of about 8 inches in depth. Suppose our trough, thus constructed, to be fixed out of doors, but under the cover of a verandah or summer-house, some few feet above the ground, on an incline of one in twenty-four, and to be surrounded with ornamental rock-work, and provided with a drain to carry off the refuse water which escapes from the lower of the two orifices; it then only remains to fix another trough on the top of the summer-house, and to connect the two tanks by a tubing of gutta-percha, passing the tube through the centre of the lower vessel, and affixing to it a nozzle of glass or ivory, the stream from which may be regulated by a stop-cock placed between the two receptacles. The salt-water, whether manufactured or taken from the sea, being thrown into the upper trough, will then descend, and, rising through the glass tube, will form a miniature *jet d'eau*, and every atom of water will thus be thoroughly aerated and purified before it reaches the anemones. The overflow may be received in a lower tank and carried upwards by hand, or by a forcing-pump, *ad infinitum*. Two cautions are necessary: one to let the cement become perfectly "dead" before the tanks are used; the other, to let all the water run off from the tank which contains the anemones about once a fortnight, when its sides and base should be thoroughly cleansed. Artificial rocks, pieces of coral and the like, may be fixed in this tank with cement as soon as it is placed in position. If the reader is fond of long words, and wishes to dignify his tank with an appropriately specific name, he may call it his "ACTINIARIUM," though this, too, is not necessary to the well-being of its contents.

The water for our tank, or glass, or pan, is the next consideration. If we live near the sea, of course fresh sea-water is advisable; let the whole, or a goodly portion, of the contents be drawn off about once a week, and a fresh supply added till the original level is regained. If the proprietor lives at a distance from the sea, and can only procure salt-water at uncertain intervals, he will find that by a continual system of aeration his sea-water will remain pure for a very considerable length of time. If he be totally unable to procure sea-water, the following recipe (which we owe to the theory and the practice of our great sea-naturalist Mr. Gosse) will answer his own purposes and those of his sea-anemones. Take

Common table-salt . .	3½ ounces.	
Epsom salts	¼ ounce.	
Chloride of magnesium	200 grains	Troy.
Chloride of potassium	40 grains	

Add spring-water (not distilled) rather less than four quarts, so that a specific-gravity bubble 1026 shall just sink in it. This solution costs about 3½*d.* a gallon. It should be filtered before it is used for the first time. When in use it should be treated in the same way as sea-water, with this exception—that a few plants of sea-weed (including coralline) should be allowed to stand in it for some days before the animals are inserted.

When evaporation takes place to any considerable extent in salt-water, it must be remembered that the *salt* does not evaporate, and therefore, unless an entirely new supply of water can be procured, spring or river, and not sea-water, must be added to the residue, or the animals will die, being unable to continue their existence in the state of *pickle* to which they have been subjected.

With regard to the *aerating*—*i. e.* the partial purifying of the water—we may adopt one or more of four methods. First, we may employ sea-weed for the purpose. Mr. Gosse recommends this proceeding in his 'Devonshire Coast;' he says that, as plants in a healthy state give out oxygen (under the influence of light) and take in carbon, and as animals reverse the process, a balance may be kept up between a due proportion of anemones and sea-weeds in any receptacle. Certain it is that sea-weeds give out oxygen under light,—the bubbles on the fronds of the *Ulva* are very striking,—and therefore they must aid in purifying the water; but, on the whole, I am inclined to think that when sea-anemones are kept for purposes of observation the sea-weeds are better omitted, for the decaying matter which they originate, and the film of green which they produce, is hardly compensated by the aerating power which they possess. For purposes of ornament, or as an interesting experiment in this curious balance of power, a few plants of green *Ulva* (*latissima* or *lactuca*) and of red *Delesseria* (*sanguinea*) or *Iridæa* (*edulis*), or of the curled and tinted Carrageen moss, may be advantageously placed among the stones of our tanks.

Another way of aerating the water is by means of the fountain already described; another, by causing water to fall in drops from one vessel into another. To effect this purpose, we may arrange our glasses in two or more rows, one above the other (say on shelves), and place in each vessel a piece of wick cotton with a long end hanging over the margin of the glass. These will work as siphons, and ensure a regular fall of water, and the overflow of the lower range may be caught in pans and carried to fill the upper glasses when empty. By this means we may imitate the flow and ebb of the tide, if we chance to think it one of the necessary conditions of our anemones' existence. But of all plans of aeration the simplest and most effectual is to use a glass syringe for the purpose, and the best syringes are those glass ear-syringes with bent tubes for self-adjustment, which one may buy at any chemist's shop. Draw off every morning a portion of the contents of each vessel into a tumbler, and then work it with the syringe till it is a tumbler of froth, and then return it to your anemones, and I am sure, by experience, that you can give them nothing else which will afford them an equal amount of pleasure. The syringe is most useful in many other ways. It serves instead of a siphon to remove water from delicate specimens; it enables you to wash the slime which adheres in rings to many of the anemones; it will remove gravel or decaying matter from the bottom of a glass without disturbing the occupants; and last, though not least, if you have a very fine "thick-skinned," and, being desirous to show his many-coloured glories to an unbelieving friend, come home and find it (as you probably would in such a case) firmly closed in a torpid state of philosophical meditation, then take my advice, and the syringe, and squirt vigorously upon his apathetic head and shoulders,—have no pity upon him, but subject him to as thorough an air and water *douche* as any Priessnitz could desire,—and if he doesn't wake up within an hour of the exhilarating application, you may throw him out of window with a safe conscience, for, in that case, you may be quite sure that he will never wake up again, being as dead as—one might almost say "as a thick-skin," for they have a most aggravating habit of dying out of hand on the smallest provocation.

A few words upon the *removal of decaying matter*, whether animal or vegetable, from the bottom and sides of our "Actiniaria," or anemone-tanks. Room, then, for his Marine Majesty's Domestic Board of Health—allow me to perform the ceremony of introduction—"Messrs. Siphon, Syringe, Gauze-net, Sponge, Glass-cloth & Co."—"The gentle and candid reader,"—and conversely. We cannot keep anemones in the perfection of health and beauty without the aid of this goodly company, and the functions of each personage are too evident to require further detail. Only, then, let me observe generally, that every tank and glass ought to be emptied at least once a fortnight (not necessarily throwing away the water, which may be aerated and returned), and then its interior is to be thoroughly cleansed by sponge and glass-cloth, with as little disturbance to the anemones which adhere to its sides as possible. I am quite aware that we may keep anemones alive for a considerable length of time without taking all this trouble,* but I am equally certain that the more often we go through this cleansing process the more vigorously will our prisoners flourish, and the more actively and beautifully will they "bloom."

On the last items, the *temperature* and the *light*, I need not enlarge. The ordinary heat of a sitting-room will effect the former condition, and for the latter a blind or shade, to ward off the summer sun, is all that is necessary. For out-of-door tanks a roof is needed to keep out the rain and protect the water from the sun-light in the summer, and if severe frosts are anticipated a more efficient shelter must be contrived, or the anemones should be removed into the house during its continuance.

If it be objected that these "conditions of existence" involve a considerable amount of trouble, I can only admit the fact, and ask whether the preservation of health—*i. e.* of the normal state of life

* Mr. Gosse has a tank containing live *Actiniæ*, &c., which has not been disturbed for more than nineteen months.

—does not always require care and foresight, especially if living objects be removed from a natural to an artificial position. Doubtless the "savage," as we call him, somewhat too contemptuously, enjoys a considerable amount of health, but when he is placed within the pale of civilization,—when, *i. e.*, he is rendered liable to the stimulus of "fire-water" and to the depressing influence of the impure exhalations of a camp or a town life,—he needs more than the former care to secure to himself those conditions which the Creator has decreed to be necessary to the well-being of this section of His creation. Evidently, then, if we have been instrumental in effecting this local and material change, it becomes our duty to secure also the additional precautions which can *only* be effected by an extra amount of trouble. And if this is true in the case of a man, it is equally true in kind, though not in degree, in the case of a SEA-ANEMONE.

CHAPTER V.

WHAT WILL IT DO WHEN I HAVE GOT IT?

NATURALISTS, it has been said, may be divided into two classes: the first comprising "collectors," or those who gather and preserve specimens, whether living or dead, for the purposes of comparison and classification; the second including the larger company of "observers," or those who combine a great love for Nature in all her forms with considerable opportunity for remarking her operations at all times and seasons, who are ever in the woods and fields, on the mountain side or by the sea-shore, and thus acquire gradually a thorough knowledge of the shapes and habits of plants and animals. But it is also said, and very justly, that a union of the methods of proceeding is necessary to ensure accuracy and certainty in any system of arrangement—that is, to constitute a perfect naturalist.

The importance of collecting for the sake of comparison is obvious. In a single species the varieties

W. Alford Lloyd del. London John Van Voorst, 1 Paternoster Row 1856 Ford & West Imp.

Actinia parasitica, "The Parasitic Anemone"

are frequently so numerous and so different in appearance that it becomes necessary to view them collectively and together with individuals of other species, in order to define their proper position and determine their relationship one to the other. Thus, too, a long series of observations on their habits, colouring, size and the like becomes equally necessary to ensure a correct judgment on the subject. For instance, some so-called species of *Actinia* have no other appreciable difference from other species than that of colour. But if it can be shown that in the same individual the colour varies at different times so greatly as to bring it within the limits of either division, or that specimens occur which combine the distinctive colouring of each assumed class, in such a case the artificial arrangement would fall to the ground.

The necessity of continual observation is strikingly shown in the case of *Anthea cereus.* We have stated that it is generically separated from the *Actiniæ*, because it does not draw back its tentacles within its body, after the fashion of the latter class. Yet I have observed a colony of these animals in a shallow pool, under the influence of a powerful summer sun, who, with few exceptions, had retracted their tentacles within the margin of their bodies till nothing but the purple tips were left exposed. Still, prolonged observation confirms the fact that it is not their habit to draw in their tentacles under the same conditions as do the anemones. In this case, when I touched these green and purple balls, instead of collapsing more closely than before, they gradually expanded into perfect bloom, whereas an anemone would shut himself up with rapid and resolute determination.

Such examples might be prolonged endlessly, yet I have said enough to show how much the humblest lover of Nature may do in the cause of Science by collecting and observing carefully, accurately and honestly, taking nothing for granted which is not proved, and (a no less useful caution) not setting down an observation as false, because he is unable to verify it immediately.

I shall now proceed to answer the question placed at the head of this Chapter, by setting down some of the curious habits and strange performances in which our friends the sea-anemones indulge, whether in captivity or in a state of freedom, remarking only, in connection with what has been just stated, that whilst some of the observations are sufficiently important to aid us in our determination of species, others again are recorded simply on account of the interest which is attached to the knowledge of the proceedings of every class of the creation.

Let us begin with the "common" anemone, *Actinia mesembryanthemum.* He is of a sluggish, apathetic, saturnine disposition, and when taken prisoner will frequently retire within himself with a dogged determination which nothing but frequent applications of fresh sea-water can remove. When in a state of freedom, unlike other species, he chooses the sides and summits of rocks which are totally deserted by the tide at low water, and remains obstinately contracted while the sun shines down upon him, and dries him up into a ball of purple slime; sometimes he is to be found in a current of running water, and then he partially opens, and appears to enjoy himself, but seems to take little trouble to place himself in such favourable situations. He will remain closed, in captivity, for days or weeks, and give no other signs of vitality beyond a regular secretion of a coat of mucus, which will need careful removal in order to maintain the purity of the water. A *douche* bath of water, and of air and water, applied by means of the syringe, will be found of great use in preserving him in a state of health and activity. It is a good plan to keep this species in a vessel by itself in order to apply these vigorous measures without disturbing the more delicate kinds.

The distinctive characters or dispositions of the various species of anemones are very curious, and well worthy of observation. It is well known that no two dogs or horses are alike, either in expression of countenance or in the character whence the expression originates. A shepherd can recognise each of his flock by its features, and in the breeding-season can tell, by the same means, to which of his ewes every new-born lamb belongs. It is said, too, that birds can be similarly distinguished. I am not prepared to state that each *individual* anemone possesses a distinguishable character of its own, but it is quite certain that each *species* has a distinctly marked set of habits, and, after a lengthened observation of our captive's ways and means of life, we are not indisposed to allow our imagination the license of asserting that a distinctive disposition is invariably included in a "specific character."

Take our friend the "daisy" (*Actinia bellis*). He is lively, sociable, easily pleased, active, amiable. As soon as we settle him in his new quarters, he will shoot out his white threads in a moment of pardonable irritation, but very quickly recovers his temper, and expands his flat disc and innumerable tentacles in a state of perpetual bloom. He moves but little, and that slowly, about his tank, but when touched contracts instantly, and presently opens himself out again, or festoons the edge of his disc in numberless graceful curves and lines of beauty. He is fond of the society of his own species, and he will congregate with his relations in a shady corner, and flourish amicably and continuously without giving his owner any anxiety about his well-doing or his prospects. I have now a most lovely group of "daisies" in a shallow glass vase, and the contrast of colour in the different varieties is very striking and beautiful.

There are several specimens of the usual description, dark gray discs of colour spotted and barred with opaque white, and occasionally enlivened with one solitary large tentacle of a rich chrome yellow. Then there comes a variety of a rich uniform chocolate, with an under colour of lurid crimson; next to him is a specimen of a delicate translucent white, perfectly ethereal in his colouring and texture; and on the other side the "queen of the daisies," of the palest, faintest amber, delicately lined with vivid vermilion, reminding me of a volcano-fire about to burst through a pavement of alabaster. In a dark room all the "daisies" *draw* towards the light, raising their bodies and becoming cup- or salver-shaped. The "cave-dweller" and the "snake-locked" anemone draw to the light in the same manner, whilst the "common" and the "thick-skinned" species are totally regardless of its influence.

The "cave-dweller" is analogous to the bat and the owl in its habits. During the day it is generally torpid, retiring, if possible, beneath the shelter of a stone, and collapsing into a pale pink or olive-green ball, but at night it wakes to life, elevates its body into a columnar form and expands its disc and tentacles. In Plate II., fig. 3, a curious phenomenon is represented, which I observed in the first individual of the species which was discovered on the North Devon coast. By the aid of a strong light and a common pocket lens a rapid circulation of water was plainly to be seen in the central hollow of each transparent tentacle. Minute black specks of an irregular oblong form were hurled along by the current, rising, eddying, falling in uncertain motion. In four out of every five of the tentacles there was a black coil of thread, sometimes apparently entire, but more frequently as though broken, which slowly circulated and occasionally remained at the summit of its receptacle, swaying to and fro in the current. When perfect these coils seemed to possess an independent motion of their own, writhing and wreathing themselves snake- or eel-wise. What these eels and black dots are is more than I can say. Mr. Gosse speaks of having seen a small Annelid in the tentacles of an anemone, but he does not think it probable that they would occur in large numbers in such a position. Further observation is required on this point, as in most others which concern this little-known class of animals. The circulation and the specks may frequently be noticed in the tentacles of the "daisy."

The "snowy" anemone agrees with the "orange-disced" and "orange-tentacled" in habit as well as in the shape and arrangement of tentacles. They all lengthen out their base into an acute oval, open freely in pure water, close rapidly when danger threatens, and are restless, changing their position frequently, and seeming to prefer the sides rather than the bottom of their tank. They are hardy and will live for a length of time in captivity. When in freedom they live in companies, and resemble the "cave-dweller" in selecting the holes bored by the Pholas, or similar cavities, as dwelling-places, and on being disturbed they draw in their tentacles and retire to the furthest limit of their dens.

The constitution of the beautiful "thick-skinned" anemones is very delicate, and they seem to take a pleasure in refuting the statement of the poet, that

"Bright things can never die,"

for, if the outer skin be ruptured, they protrude their "convoluted bands," which presently mortify and slough away, whereupon the animal dwindles and decreases without ceremony. If they be taken off their native rocks without injury they will flourish and bloom for a considerable time. They are inactive, and move but little upon their bases; if they want to make a journey they detach themselves from their moorings, inflate their bodies with water, draw in their tentacles, and surrender themselves to the guidance of the tide and the currents. Thus we generally find them in companies at the further extremity of deep chasms in the rocks, whither they have been washed, and where they adhere and expand their tentacles in quest of food.

The "gem" is fickle and lethargic. He shuts himself up as if he wanted the colouring of his speckled body to be noticed and admired, though the markings of his disc are no less beautiful. He takes no trouble about fixing himself in new quarters, but lies indolently on his side, and lazily puts out a tentacle or two in case any tit-bit should happen to float by within his reach.

One word about *Anthea cereus*, our eccentric acquaintance "Legs." I am inclined to think that he possesses a philosophic mind of a very high order. He is always contented and cheerful and active, and makes the best of whatever situation he finds himself placed in. As soon as he is thrown into a glass or tank he lies sprawling and kicking about at the bottom for a short time, apparently in a state of no small bewilderment, but quickly recovers himself, and climbs rapidly up the sides of the receptacle, where he fastens himself just under the water-line, so that he can sweep the under surface of the air with his long tentacles, and these he is always waving about in all directions with a graceful sweeping or a quick jerking motion, so that a colony of these animals in full health resembles a forest of poplars tossing and heaving in a whirlwind.

Let me, in conclusion, again urge the great importance of careful and accurate observation, both of habits and colouring, as well as of the "habitats," or the places and positions in which the sea-anemones are to be found.

If the observer have no time or inclination to

apply the results which he has gained [illegible] the knotty questions which abound in [illegible] he may yet materially advance the [illegible] History, if he will keep a written [illegible] has found and seen, and place it [illegible] in the hands of those who are [illegible] many difficulties which surround [illegible] specimen of such a record is given [illegible] The observations should be [illegible] accompanied with coloured [illegible] attention should be paid to facts [illegible] settle the question of "species" [illegible] of the use of the so-called "[illegible]" [illegible] "spike-cases."

And I may add, that if [illegible] to take delight in the study [illegible] he will do well to show [illegible] which leads to this [illegible] pleasure, and usefulness, [illegible] he may be sure that the [illegible] the more will he recognise [illegible] Nature, and the more will [illegible] forming the duties which [illegible] his own station in the [illegible] like himself, are of a twofold [illegible] spiritual. He will be convinced [illegible] of body or mind, is Nature, [illegible] butive consequence of [illegible] Nature. He will eventually [illegible] gians' "sin" and the political-economists' "crime" are as much diseases and as curable as the medical man's "cholera" and "typhus." He will further get to realise the old proverb that "Prevention is better than cure," nay, that (taking man as a whole) it is the only true cure. He will thus, if he be more than a word-maker and a theorist, strive to place his own soul and body, and those of his brethren, within the influence of those conditions of existence which only are true because only from God.

He will avoid mental and bodily infection, contagion, malaria and the like; nay, he will avoid medicines which cure by exciting disease, for when he has once arrived at a conclusion on the subject of the laws of health, he will know that, if they be unbroken, neither disease nor medicine—its consequence—need ever be incurred. He will be content if he can attain his "summum bonum," the "mens sana in corpore sano;" he cannot be too thankful if he be allowed upon earth to exercise perfectly his perfect faculties in strains which may thus be best trained for the choruses of eternity.

In a word, then, he will ensure to himself, his fellow-man, and all created objects with which he is in relation, those conditions of healthy existence which have been constituted such by an all-wise Creator; he will thus apply all hisenergies in promoting the spiritual well-doing and well-being of the whole race of mankind, and in glorifying the God of the natural and the invisible world.

The foregoing pages are a contribution (however unworthy it be) to the great cause, and it is an unspeakable pleasure and privilege to be able to add even a single stone to that great pyramid of Inductive Science, whose base extends over the whole earth, and whose summit is now hidden, but shall hereafter be revealed in the Heavens. In this manner the humblest and most unlearned mind, provided only it love Nature honestly and with all its powers, may join in the prophetic strain of the great lyric poet of the Latin world, and (though it place a construction on the poet's words which he did not contemplate) may yet sing perpetually and thankfully in the great chorus of its fellow-workers :—

> "Exegi monumentum ære perennius,
> Regalique situ pyramidum altius,
> Quod non imber edax, non Aquila impotens,
> Possit diruere, aut innumerabilis
> Annorum series, aut fuga temporum."

CHAPTER VI.

SUPPLEMENTARY AND CRITICAL.

ON THE DISTINCTIONS OF GENERA AND SPECIES.

In the third chapter of this 'Manual' I have endeavoured to popularise a new arrangement of the *Actiniæ*.

Whether the scheme be correct or no may well be an open question, and one which can only be solved by repeated and accurate observation. There can be, however, no doubt that some such a system is greatly needed. Naturalists in bygone days greatly overlooked the lower orders of creation, and though, in the present day, many beautiful monographs exist to confute a similar charge of ignorance or neglect, it is still true that no complete account of the family of the *Actiniæ* has ever been compiled. Perhaps it is not too much to assert that no naturalist has ever yet paid sufficient attention to the subject to warrant him in giving a monograph of the *Actiniæ* to the world. There are many points of classification and of minute anatomy which are vexed questions, and it is very probable that a careful examination of our shores would largely add to the recorded number of species and varieties.

I have read, I think, all that has been written on the subject; I have collected for myself and seen the collections of others; and the natural result is, that I have settled in my own mind many questions which arose as to the relative position of various individuals of the family.

Such conclusions are submitted to the naturalist in the tabular arrangement which ends this Chapter, and I shall now venture to add a few of the reasons which have influenced my decision, and to hope that these imperfect thoughts may be not altogether untrue or entirely useless in aiding some future observer to build up a more perfect system of natural arrangement.

The first question which we have to answer is the following:—

"What is a Genus?"

"A genus," says Professor Balfour, "is an assemblage of nearly related species, agreeing with one another in general structure and appearance more closely than they accord with any other species."

Now Mr. Gosse perceived that the known species of sea-anemones were naturally divided into three such assemblages, and he accordingly divided the recognized genus *Actinia* into three genera, to which the three groups I have given in the text correspond, and I only did not use his nomenclature because it involved the continual repetition of two more "hard words," a great consideration in a *popular* 'Manual.' I venture to copy his arrangement from the first volume of his 'Manual of Marine Zoology for the British Isles,' a work which should be in the hands of every shore-going naturalist.

"*Actinia* (Linnæus). Body adherent, cylindrical; destitute of warts, of pores, and of missile filaments; skin smooth; a series of capsuliferous spherules at the margin of the disk.

A. mesembryanthemum.
,, margaritifera.
,, chiococca."

This genus corresponds with Group I. as given above. By "missile filaments" are meant the white threads which the "daisy" and his allies send out on provocation. By "warts" those perforated erect tubercles which stand out from the whole exterior surface of the "thick-skinned" and his friends, and by "pores" those less apparent projections which line the upper portion of the body of the "daisy" and company. Mr. Gosse seems thus to separate specifically the tubercles of either genus, and to believe that the "sucking-glands" are distinct from both these organs and are clearly marked in both genera. But all my observation goes to prove that the genera *Sagartia* and *Bunodes* have tubercles in common, agreeing in their use as sucking-glands, and disagreeing only in size and distribution. I can also prove that *Sagartia* possesses certain *invisible* pores or channels through which the missile filaments are expelled and retracted.

By capsuliferous spherules are meant the blue beads which are inside the upper margin of the body, and which contain the capsules or spike-cases spoken of in Chapter III.

"*Bunodes* (Gosse). Body adherent, cylindrical, studded with warts; skin leathery; not emitting missile filaments—nettling threads long and simple; tentacles generally thick, conical, obtuse.

B. gemmacea.
„ thallia.
„ clavata.
„ crassicornis."

This genus corresponds with Group II.

"Nettling-threads" are the "spikes" mentioned in Chapter III., and figured, together with their cases, in Plate II., fig. 1. *B. crassicornis* is the "thick-skinned" of the text, which I have ventured to designate *coriacea* for reasons there given.

"*Sagartia* (Gosse). Body adherent, cylindrical, without a skin; destitute of warts; emitting capsuliferous filaments from pores; nettling-threads short, densely armed with a brush of hairs; tentacles conical.

S. viduata = anguicoma.
„ troglodytes.
„ aurora.
„ candida.
„ rosea.
„ nivea.
„ venusta.
„ parasitica.
„ bellis.
„ dianthus."

Corresponding with Group III.

I am at a loss to know what is here meant by the statement that this group is "without a skin," unless the phrase allude to the "*lobed* skin" which the author describes (four lines above this remark) as being a characteristic of the genus *Capnea.* As I have said, I believe that this genus *Sagartia* possesses warts exactly similar to those of *Bunodes* in all respects except size and distribution; and I also believe that the pores from which the threads proceed are not the warts to which I allude. By "capsuliferous filaments" are meant the missile filaments previously spoken of, and they are called capsuliferous, because they bear the capsules or spike-cases. The nettling-threads or spikes are stated to be densely armed with a brush of hairs: this is not represented in Plate II., because it is only to be observed by a much higher microscopic power than will serve to show the spikes themselves with sufficient clearness.

So far of genera. Our next question really claims a priority in point of time and importance, but, for obvious reasons it comes "last, though not least."

What are the characteristic marks which warrant us in defining a species?

I know of few questions so difficult to answer as this—

"What is a Species?"

"By species," says Professor Balfour, "are meant so many individuals as are presumed to have been formed at the creation of the world, and to have been perpetuated ever since."

These, when *temporarily* influenced in form, size or colour by climate, position or nourishment, become varieties.

"A variety has a constant tendency to revert to the original type." A "permanent" variety or race may be induced by constant artificial stimulus; it has still a tendency, though in a lesser degree, to return to the type.

Now I have collected (see Chapter V.) four or more individual sea-anemones which differ entirely from one another in colour, and in this exclusively:

I call them varieties of *Actinia bellis.* I collect a group of animals (see Chapter III.) similarly differing among themselves, and call them varieties of *A. mesembryanthemum.* Also another group (same Chapter) varieties of *A. coriacea.* Lastly a group (same Chapter), whose only difference is one of colour also, which I am directed to divide into three *species*, and call them respectively *Actinia nivea, venusta* and *aurora.*

Why should I act in this inconsistent manner?

That I cannot answer.

What reason have I for asserting that colour is the distinctive mark of a variety, and form and habit that of a species?

The fact is that we know next to nothing about the state of sea-anemones at the creation.

If we take a plant, and propagate it by seed for several years, and it continue to bring forth similar plants, in proportion to the length of time in which it remains constant will be our certainty of its being a species. If it alter, its former state will be the form of a variety, and the alteration will be that of the species, probably; and a number of similar plants similarly altering will confirm our opinion.

So, I cannot see any method of absolutely ensuring the truth as to the species of sea-anemones, except a like method of careful breeding and accurate observation.

But in the mean time what are we to say?

I am rather ashamed of writing anything which is so loosely determinative of the question, but I do write it because no one else has written anything which is even equally satisfactory, at least to my mind.

Let us be at any rate consistent. If colour be admitted to be characteristic of species, then let us allow the "tiger" and the "white-tentacled thick-skinned" to be as much species as the "snowy" and the "orange-disced" anemones, but if identity of form be considered similarly characteristic let us then speak of all those anemones which agree in form as individuals of a species, and make the mere distinction of colour to constitute, uniformly, varieties.

If a "strawberry" would only be good enough to breed a "tiger," or a "snowy" an "orange-disced" anemone, or if they would be sufficiently considerate to turn one into the other in our anemone-tanks, we might have better data to reason upon. But as things are, we only generalise a type from the various individuals which are presented to our notice; and in a given number of such individuals SHAPE and HABIT do seem to be so distinctly recognisable and so definitely limited, and COLOUR to be so varying, variable and interchangeable in objects otherwise apparently identical, that, until further observation be recorded, I cannot help believing that the former may be characteristic of *species*, and the latter of *variety*.

With these opinions I have drawn up and now subjoin a tabular arrangement of species and varieties, which differs in many respects from the recorded opinions of naturalists on the subject.

One point alone remains to be explained. The "strawberry anemone" is here set down as a species; and a casual observer will remark that its difference from *A. mesembryanthemum* is only one of colour, and that therefore I am erring against my own theory. But the fact is that there is a difference of form; the largest "strawberry" is twice the size of the largest *mesembryanthemum* which can be found, and on this ground I have made the separation.

I have now only to add that I trust the reader will understand that when I differ from other observers I do so in no dogmatic or presumptuous spirit, but simply from the fact that I have acted in accordance with what I believe to be the practice of all honest naturalists. I have given considerable time and attention to the subject, and have stated fairly and accurately all that I have seen, and the conclusion which my own reason has led me to infer from my own observations and those of others, whether recorded orally or in books.

I would certainly not be dogmatic, for I believe that the most learned in sea-anemones (as possibly in many other subjects), is consequently only the more convinced that he knows but little, and is considerably at a loss to perceive how he may best apply and arrange the little that he knows.

TABULAR ARRANGEMENT OF THE GENERA, SPECIES AND VARIETIES OF THE ACTINIÆ.

Family.	Genera.	Generic difference.	Species.	Specific difference.	Varieties.	Var. difference.
ACTINIADÆ.	I. ACTINIA.	Smooth-skinned: row of beads on oral disc, close to margin of body, containing spike-cases.	1. *mesembryanthemum.*	Skin soft.	*a.* fusca.	Brown.
					b. olivacea.	Olive-green.
					c. striata.	Apple-green, with pea-green stripes.
					d. prasina.	Leek-green.
					e. hepatica.	Liver-colour.
					f. rufa.	Light red.
					g. tigrina.	Scarlet-red, with yellow stripes.
					h. aurantia.	Orange-buff.
					i. chiococca.	Beads white.
			2. *fragacea.*	Ditto; considerably larger, spotted.		
			3. *margaritifera.*	Skin leathery.		
	II. BUNODES.	Skin covered with numerous conical, coloured, perforated warts.	1. *gemmacea.*	Warts arranged in vertical rows, six or more of which are larger than the rest.	*a.* vulgaris.	
					b. thallia.	Disc, yellow rays on black ground.
			2. *clavata.*	Warts arranged in vertical rows, all equal; tentacles few, marginal. Found in crevices, Pholas holes, &c.	*a.* aurantia.	Yellow.
					b. rosea.	Rose.
					c. glauca.	Glaucous.
			3. *coriacea.*	Warts irregularly arranged.	Varieties numberless.	
	III. SAGARTIA.	Warts flattened, and on upper surface of body only; spike-cases in white threads.	1. *bellis.*	Tentacles horizontal.	*a.* vulgaris.	
					b. cornuta.	One yellow tentacle in the disc.
					c. eburnea.	Disc ivory-white.
					d. tyriensis.	Disc chocolate and crimson.
					e. Regalis.	Disc alabaster and vermilion.
			2. *troglodytes.*	Tentacles vertical. Habit, hiding under stones, &c.		

		3. *anguicoma.*	Tentacles marginal, few, long, taper, flexible. Habit, alternately flat and columnar.		
		4. *candida.*	Tentacles graduated; size small (⅛-inch).	*a.* vulgaris. *b.* immaculata.	 No brown ring round the tentacle-bases.
		5. *dianthus.*	Margin of disc fringed and much lobed.	*a.* alba. *b.* lactea. *c.* carnea. *d.* rufa. *e.* aurantia. *f.* olivacea.	White. Cream-coloured. Flesh-coloured. Pale red. Orange. Olive.
		6. *parasitica.*	Outer row of tentacles smaller than others, and everted; body columnar, of large size. Habit, parasitic.		
		7. *aurora.*	Outer row of tentacles smaller than others, and everted; tentacles about 200, graduated, erect, short. Body opaque, conical.	*a.* "venusta." *b.* annulata. *c.* "rosea." *d.* "nivea." *e.* Devoniensis. *f.* "aurora." *g.* subrufa.	Body and disc orange, tentacles white. Body and disc orange, tentacles white with black ring round their bases. Body orange-brown, tentacles rose. Body olive, disc and tentacles white. Body olive, tentacles lilac. Body olive, disc green, spotted with white, tentacles orange. Body olive, tentacles russet and striped.
		a. alba. *b.* chrysanthellum. *c.* biserialis. *d.* vermicularis. *e.* intestinalis.	Not placed, as not having been sufficiently described by their discoverers.		

REMARKS ON A FEW OF THE RARER SPECIES AND VARIETIES DESCRIBED IN THE FOREGOING TABLE.

A. clavata, "The Weymouth Anemone" (Thompson).—This anemone was discovered in 1851 by W. Thompson, Esq., the eminent naturalist and photographist, of Weymouth, to whose well-known courtesy I owe the following interesting observations on its form, habits and habitats:—

"Body cylindrical, ¼ inch in diameter; tentacula placed in two series, one being much longer than the other, club-shaped, larger at the top than the bottom, and ending abruptly. Twenty-five longitudinal raised lines are placed at regular intervals round the body, the top of each produced into a wart at the edge of the disc. When expanded, measures 2½ inches from tip to tip of the tentacles. Skin warty, ground-colour straw or yellowish or rosy pink, profusely covered with innumerable small puce-coloured specks, which become scarcer towards the apex, where they form five or six circles; the raised longitudinal lines have fewer spots on them than the rest of the body. The shorter tentacles are of a uniform dirty transparent white, the longer have blotches of pink or puce upon the white ground. In the interior of the tentacles are visible small oval cream-coloured bodies. Oral disc pellucid, with the appearance of having been sprinkled with chalk.

"It inhabits rocky ridges at extreme low-water mark, taking possession of the deserted hole of a Pholas, or some other rock-boring mollusc; it is also partial to narrow crevices. When alarmed it withdraws itself completely within its hole. This species is now very rare, in consequence of its being much sought after from its great beauty in an aquarium. From the different spots which it selects for a home many more are destroyed by ignorant curiosity-hunters than are secured alive. It is described in the 'Zoologist' for 1851, Appendix cxxvii."

A. thallia, "The Glaucous Anemone" (Gosse).—This anemone is described in Mr. Gosse's 'Tenby' as a new species. It differs from the "Gem" in colour and in its "superior size." But if my theory be right that colour is a mark of variety, and not of species, we are thrown back on the size of the animal as the only distinctive specific mark. Mr. Gosse states that its disc is 2 inches in diameter. Many specimens, however, of the ordinary "Gem" have been brought to me from Morte-stone of a similar size; one, in particular, measuring 2½ inches across. I therefore conclude *thallia* to be only a colour variety of *A. gemmacea.*

A. anguicoma, "The Snake-locked Anemone."—Mr. Gosse makes the *A. anguicoma* of Prof. Johnston equal the *A. viduata* of the same author. He further describes and depicts (in his 'Devonshire Coast') as an *anguicoma* that which I consider to be a *troglodytes.* Undoubted specimens of *troglodytes* have the vertical tentacle streaks which are given by Johnston as characteristic of *viduata.* It is probable that these two species are identical, and that *anguicoma* (distinguished by its long and slender tentacles, and its power of elongating itself from ½ an inch to *five inches and a half,* Johnston) is unknown on the Devonshire coast.

A. dianthus, "The Plumed Anemone."—Mr. W. Thompson informs me that this is essentially a deep-water species at Weymouth, and is never taken in less than three fathoms water. *Small* specimens have been forwarded to me by Mr. G. H. Lewes from Tenby, where he finds them on the rocks at extreme low water. One small specimen has been discovered on Morte-stone, near Ilfracombe, by my friend Mr. W. Brodrick. Mr. Thompson writes:—"The assertion that *A. dianthus* cannot be detached without injury to its base is a mistake; they are not more difficult to detach than are other species, and move constantly of their own accord from place to place. From what I have seen of *A. dianthus*, and some hundreds have passed through my hands, I am of the same opinion as our lamented friend Dr. Johnston, that Mr. Couch's *A. dianthus*, which he finds between tide-marks, is a distinct species."

A. parasitica, "The Parasitic Anemone." — On this species Mr. Thompson writes to me as follows:— "It is by no means scarce in Weymouth Bay; I dredge two or three of a day, and after a storm they may be found thrown on the shore. I have never found them on bivalves or *living* univalves: they are occasionally taken on stones. Their favourite site is in a shell of the common whelk, *B. undatum*, which is inhabited by a hermit-crab. The crab walks about, if not unconscious of, at all events not caring for, its living load, and the *Actinia* no doubt rides about from one feeding-ground to another, with great advantage to itself, if not with great pleasure."

A. aurora.—It will be seen that I have grouped four of Mr. Gosse's species and two new varieties (which have at least equal claims to be considered as species) under one head. During the last two years I have seen numberless specimens of all these kinds from Tenby, from Lundy Island, and from Morte-stone, and there is no doubt in my own mind that, however they may differ in colour, and even in the number of tentacles (arising from their difference of *age*), they may yet be classed under one typical *form*, which I have described in the foregoing Table as characteristic of the species.

It is only by collecting and comparing vast numbers of *Actiniæ* that we can arrive at a trustworthy generalization with regard to the characteristics of genera, species and varieties.

APPENDIX I.

SPECIMEN OF A NATURALIST'S DIARY.

(See page 85.)

N.B. The first four items may be kept on the left-hand side of the book, and the opposite pages be reserved for the record of events. An Appendix should contain more elaborate notices of microscopic observations and the like.

Group II.

Name.	Where found.	When.	Under what circumstances.	Record of events in the animal's prison-life. (With dates.)
A. gemmacea "The Gem."	Torr Point, Ilfracombe	Aug. 23, 1855	Extreme low-water, under a stone.	Aug. 24. Fixed firmly to side of tank. 25. Expanded. 28. Becomes paler in colour externally. 29. In full health, but lazy; gave him a douche bath. Same day, evening. Lively.
A. coriacea "Thick-skinned."	Rat Island, Lundy	Aug. 27, 1855	About half-tide. In a rock-pool, under water, expanded.	Aug. 28. In full bloom (var. white-tentacled). 29. Sends out the "ovaries" from a wound in the base. 30. The green colour of the body quite faded away. Sept. 2. The "ovaries" slough. 3. He dies.

Group III.

Name.	Where found.	When.	Under what circumstances.	Record of events in the animal's prison-life. (With dates.)
A. bellis "The Daisy."	Lee, Ilfracombe	Aug. 29, 1855	In a rock-crevice, a few yards below high-water mark (neap-tide).	Aug. 30. Blooms. 31. Walks up the sides of the glass, and expands his body into a salver-like shape. Sept. 2. Congregates with his fellows. 3. Examined the circulation in the tentacles, and the structure of the "threads" with the microscope. (See Appendix; p. —.) 4. His disc circular and flat. 5. Disc festooned.

APPENDIX II.

HABITATS OF ACTINIÆ.

THE following list is appended, in order to give the wandering naturalist some idea of the sea-anemones which he is likely to encounter at various stages of his travels. It is naturally imperfect, since but little attention has hitherto been paid to the subject, and the "smallest contribution" to its future efficiency will be "thankfully received" by the compiler.

The last edition of Johnston's 'Zoophytes,' Mr. Gosse's publications, and personal observation, have been the only available sources of information.

The names which follow each habitat are those of the first observers of the fact recorded.

1. *Actinia mesembryanthemum.* Universally distributed.
2. *A. mesembryanthemum,* var. *aurantia.* Lundy Island, 1855 (the Author).

3. *A. mesembryanthemum*, var. *chiococca*. Coast of Cornwall, especially St. Ives (W. Cocks).
4. *A. fragacea (A. mesembryanthemum)*. Universally distributed.
5. *A. margaritifera*. Copeland Isle, 1811 (J. Templeton); Donegal Bay (E. Forbes).
6. *A. gemmacea*. Coast of Cornwall (Gærtner); Ilfracombe (Ralfs); Lundy (the Author).
7. .*A gemmacea*, var. *thallia (A. thallia)*. Lidstep, Pembrokeshire (Gosse).
8. *A. coriacea (A. crassicornis)*. Very generally distributed.
9. *A. clavata (A. clavata)*. Weymouth (Mr. W. Thompson).
10. *A. bellis*. Cornwall (Gærtner); Isle of Rathlin, 1795 (Templeton); Ballyhome Bay, County Down (Thompson); Oddicombe, South Devon —Ilfracombe (Gosse); Lundy (the Author); Weymouth (Mr. W. Thompson).
11. *A. bellis*, var. *eburnea*. Ilfracombe (the Author).
12. *A. bellis*, var. *tyriensis*. Ilfracombe (Gosse; the Author).
13. *A. bellis*, var. *regalis*. Ilfracombe (the Author).
14. *A. troglodytes*. Berwick Bay (Johnston); Cornwall (Couch); Isle of Man (Forbes); Moray Frith (Robertson); Tenby (Gosse); Ilfracombe (the Author).
15. *A. anguicoma*. Bangor (Price); Clare (Thompson); Ilfracombe (?) St. Mary Church (Gosse).

16. *A. dianthus (A. dianthus* and *A. plumosa).* Johnston gives no specific habitat; Cornwall (Couch); Weymouth (Mr. W. Thompson); Isle of Ambrae (Landsborough); Tenby (Mr. G. H. Lewes); Torquay, Morte-stone (Mr. Brodrick).
17. *A. parasitica.* Cornwall (Couch); Weymouth (Gosse).
18. *A. rosea.* Babbicombe, South Devon (Gosse).
19. *A. aurora.* Tenby (Gosse); Lundy, Morte-stone (the Author).
20. *A. venusta.* Tenby (Gosse); Lundy, Morte-stone (the Author).
21. *A. venusta,* var. *nivea (A. nivea).* Babbicome, Ilfracombe (Gosse); Lundy, Morte-stone (the Author); Tenby (Mr. G. H. Lewes).
22. *A. candida.* Ilfracombe (Gosse).
23. *A. candida,* var. *immaculata.* Ilfracombe (the Author).

The following anemones have not been sufficiently described to warrant us in attaching them to either of the above genera:—

24. *Actinia alba.* Cornwall (Cocks).
25. *A. chrysanthellum.* Cornwall (Couch).
26. *A. biserialis.* Guernsey (Forbes).
27. *A. intestinalis.* Shetland (Fleming).
28. *A. vermicularis.* Shetland (Forbes).

APPENDIX III.

GLOSSARY OF "HARD WORDS."

It was said of Rienzi, "the Last of the Tribunes," and it is no small praise, that "he loved to dispense his knowledge in familiar language." We have seen that it is impossible to avoid using words compounded from the dead languages, when writing on Natural History subjects, but at the same time it is very possible to translate and explain them, in order that we ourselves and others may use them intelligently, and remember them easily.

Wherefore I compose a Glossary,* as a necessary addition to my book, and herewith (having

* This Glossary comprises all the "hard words" which belong to the subject of anemones, and which are to be found in the six Chapters of this 'Manual.'

said all which at present I have to say upon the subject) do conclude my labours, and wish that desirable unit, my courteous reader, most heartily *farewell.*

Acalephs, "stinging" animals, *e. g.* jelly-fish (*acaléphé*, a nettle, Greek).

Actinia, a sea-anemone (*actis*, a ray of the sun, Greek; in allusion to the circular arrangement of the tentacles.

Anguicoma, the specific name of the snake-locked anemone (*Anguis*, snake; *coma*, hair, Latin).

Annelids, animals whose bodies are formed of rings, *e.g.* earth-worms (*annulus*, or *anulus*, a ring, Latin).

Anthea, the next of kin to the anemones (*anthos*, a flower, Greek).

Aquarium, a tank or vessel of water (*aqua*, water, Latin).

Articulate, possessing a jointed external skeleton (*articulus*, a joint, Latin).

Asteroid, star-shaped (*aster*, a star; *eidos*, shape, Greek).

Aurora, the specific name of the orange-tentacled anemone (*aurora*, the dawn, Latin; perhaps an allusion to its colour).

Bellis, the specific name of the daisy anemone (*bellis*, a daisy, Latin).

Bunodes, one of Mr. Gosse's new genera — Chapter VI. (*bounos*, a hill; *eidos*, shape; alluding to the outstanding warts which characterize the genus).

Candida, the specific name of a white anemone (*candidus*, white, Latin).

Capsule, a small receptacle (*capsula*, a little chest, Latin).

Cereus, the specific name of the common Anthea (*cereus*, waxen, Latin).

Chrysanthemum, the flower so called (*chrysos*, gold; *anthos*, a flower, Greek).

Cilia, minute hairs attached to various parts of the bodies of animals, and which vibrate rapidly (*cilium*, an eyelash, Latin).

Coriacea, the specific name of the thick-skinned anemone (*corium*, Latin; *chorion*, Greek, skin).

Dianthus, the specific name of the plumed anemone (*anthos*, the flower; *Dios*, of Jupiter).

[*Plumosa*, synonym; *pluma*, a plume, Latin.]

Disc, the exterior portion of an anemone included by the upper margin of the body (*diskos*, a quoit, a round flat dish, Greek).

Echinoderm, a prickly-skinned animal, *e. g.* star-fish (*derma*, skin; *echinos*, sea-urchin; *echein*, to hold fast, grasp, Greek).

Exegi monumentum, &c., the lines at the conclusion of Chapter V., may be thus (somewhat freely) translated:—

Yes! I have raised a fane more durable than brass,
And higher than the height of kingly pyramids,
Which not the washing rain, or north wind impotent,
May hurl assunder; nor the innumerable course
Of years, or flight of ages e'er shall dissipate.

Fane, a temple (*fanum*, a consecrated building or portion of land, Latin).

Gelatinous, jelly-like (gelatine).

Gemmacea, the specific name of the Gem-anemone (*gemma*, a gem, precious stone, Latin).

Genus, a collection of species which agree among themselves more closely than they coincide with other species.

Habitat, the place in which animals or plants are found (*habitare*, to inhabit, Latin).

Helianthoid, sunflower-shaped (*Helios*, the sun; *anthos*, a flower; *eidos*, shape, Greek).

Hydraform, hydra-like (*hydra*, an animal which lives in the water; *hudor*, water, Greek; *forms*, a shape, Latin).*

* The Hydra of old story was a monster who lived in some marshes, and destroyed people. As soon as one head was cut off, two more appeared, unless the actual cautery was applied. He was probably a malaria, and the moral of the allegory is that half-sanitory measures are worse than none, since they excite and do not remove the causes of the evil. The hydra *viridis* is a freshwater zoophyte, bearing many polyps on a green stem. He lives in marshes, and reproduces himself even faster than his prototype.

Inductive, induction is the act of arriving at the perception of a law of creation, from the examination and comparison of many objects in which the law is exercised.

Lucernaria, a "lamp-like" zoophyte (*lucerna*, a lamp; *lux*, light, Latin).

Margaritifera, the specific name of the pearly anemone (*margarita*, a pearl; *ferre*, to bear, Latin).

Margin, edge or boundary line (*margo*, brink, verge, Latin).

Mesembryanthemum, the specific name of the common ane-anemone (perhaps from *mesos*, middle; *hemera*, day; *anthos*, flower, Greek, "The mid-day flower" or "flower of noon").

Mollusc, an animal with a soft body, *i. e.* not possessing an internal or external skeleton—and not a radiate (*molluscus*, soft, Latin).

Monograph, an essay on a single tribe, or genus, or species of animals or plants (*monos*, single; *graphein*, to write, Greek).

Nivea, the specific name of the snowy anemone (*nix*, *nivis*, snow, Latin).

Normal, in accordance with the law of creation, *i. e.* an instance of a rule, and not its exception (*norma*, a law. Latin).

Organic, possessing organs, *e. g.* stomach, lungs, &c., parts whose office it is to gather and appropriate nourishment from surrounding objects, and otherwise carry on existence.

Ovaries, the organs whence the eggs (*ovum*, an egg, Latin) originate.

Parasitic, the specific name of an anemone. A parasite was a person who stationed himself (*para*) beside (*siton*) the food (Greek) of another; hence, originally, a messmate, &c., then it degenerated (together with the practice) into the idea of flattering, "toadying," and the like; afterwards it became applied in Natural History to those animals who took up their abode on or about some other animals or some plants, and did not select a freehold of their own.

Physiology, the history of natural objects, more generally applied to organic objects (*physis*, nature; *logos*, a word, Greek).

Platycheles, the specific name of the "broad-clawed" crab (*platys*, broad; *chelai*, claws, Greek).

Polyp, many-footed, the synonyme of the true zoophyte: in allusion to the tentacles of the animals (*polus*, many; *pous*, foot, Greek).

Radiate, possessing organs arranged in rays round a centre (*radius*, a ray, Latin).

Rosea, the specific name of the rosy anemone (*rosa*, a rose, Latin).

Sagartia, one of Mr. Gosse's new genera: see Chapter VI. (perhaps from *sagé*, armour, and *arteesthai* (Hdt.), to make ready, Greek; in allusion to the missile "threads" which characterize the genus).

Scylla and Charybdis. At Charybdis, now Kalofaro, in Sicily, are numerous and variable sea-currents; and on the opposite coast of Calabria is a perpendicular rock, honeycombed by the sea into numberless caverns, which is called Scylla. To avoid the dangers of Charybdis the ancient mariners were wont to draw too near Scylla, and thus provoked the shipwreck which ensued. Hence the old proverb,

> "Who flies Charybdis, upon Scylla strikes."

Species, a collection of individuals, whose present form has been perpetuated without change since the creation.

Synonym, another name for the same thing (*sün*, together with; *onoma*, a name, Greek).

Tentacles, organs which grasp (*tentare*, to touch, handle, attack, Latin).

Thallia, a specific name of the glaucous anemone (*thallos*, a young shoot, especially of an olive, Greek). [*Glaucous* means grey-blue, and alludes to the colour of the olive.]

Troglodytes, the specific name of the cave-dweller (*troglé*, a cave; *duein*, to creep into, Greek).

Vasculum, the term applied to portable cases for collecting plants and animals (*vasculum*, a little vessel or receptacle; *vas*, a vessel, Latin).

Venusta, a specific name of the orange-disced anemone (*venustus*, beautiful, Latin).

Vertebrate, possessing a series of vertebræ, or hollow bony rings, which, when united, form a back-bone, as in man, fish.

Vivarium, an enclosure or receptacle for live animals (*vivus*, living, Latin)

Zoanthus, "live-flower" (*zoos*, living; *anthos*, flower, Greek).

Zoology, the natural history of animals, or the science of animal life (*zoé*, life; *logos*, a word, Greek).

Zoophyte, literally, an "animal-plant" (*zoon*, an animal; *phyton*, a plant, Greek).

NOTE.

Whilst this volume has been passing through the press, the following interesting fact, with reference to the duration of life among the *Actinidæ*, has been kindly communicated to me by Mr. Van Voorst. He says: "My friend the Rev. Dr. Pollexfen, who is now here, tells me that he saw a few days ago, in the possession of Professor Fleming at Edinburgh, an *Actinia mesembryanthemum*, taken at North Berwick, in 1828 (and at that time imagined to be seven years old), by the late Sir John Dalyell, who kept it in confinement till his decease, and in whose work, 'Rare and Remarkable Animals of Scotland,' published in 1848, there is a figure of it. This animal must, therefore, be about thirty-five years old, twenty-eight years of which it has sustained life in confinement."

INDEX OF TABLES.

GENERAL INDEX.

ERRATA.

Page 5, line 3 from bottom, *between* our *and* friend, *insert* respondent

Page 12, line 13, *after* continued, *insert* as the spinal cord or "marrow"

Page 13, line 9, *dele* Latin

Page 26, line 17, *for* 11 *read* II.

Page 27, line 17, *for* 11 *read* II.

Pages 54—57. This arrangement of the new genus *Sagartia* is in accordance with the ordinarily received notions on the subject. The author's own classification may be seen in pages 98, 99.

CPSIA information can be obtained at www.ICGtesting.com
Printed in the USA
BVOW071538171011

273850BV00009B/169/P